MATHEMATICS AND POLITICS

MATHEMATICS AND POLITICS

Strategy, Voting, Power and Proof

Second Edition

Alan D. Taylor

Department of Mathematics
Union College ■ Schenectady, NY

and

Allison M. Pacelli

Department of Mathematics
Williams College ■ Williamstown, MA

 Springer

Alan D. Taylor
Union College
Schenectady, NY, USA

Allison M. Pacelli
Williams College
Williamstown, MA, USA

ISBN: 978-1-4419-2661-6   e-ISBN: 978-0-387-77645-3
DOI: 10.1007/978-0-387-77645-3

Printed on acid-free paper

springer.com

Allison Pacelli dedicates this book to her mother Patricia Pacelli.
Alan Taylor once again dedicates it to Gwendolyn and Harrison.

Preface

Why would anyone bid \$3.25 in an auction where the prize is a single dollar bill? Can one "game" explain the apparent irrationality behind both the arms race of the 1980s and the libretto of Puccini's opera *Tosca*? How can one calculation suggest the president has 4 percent of the power in the United States federal system while another suggests that he or she controls 77 percent? Is democracy (in the sense of reflecting the will of the people) impossible?

Questions like these quite surprisingly provide a very nice forum for some fundamental mathematical activities: symbolic representation and manipulation, model–theoretic analysis, quantitative representation and calculation, and deduction as embodied in the presentation of mathematical proof as convincing argument. We believe that an exposure to aspects of mathematics such as these should be an integral part of a liberal arts education. Our hope is that this book will serve as a text for freshman-sophomore level courses, aimed primarily at students in the humanities and social sciences, that will provide this sort of exposure. A number of colleges and universities already have interdisciplinary freshman seminars where this could take place.

Most mathematics texts for nonscience majors try to show that mathematics can be applied to many different disciplines. A student's interest in a particular application, however, often depends on his or her general interest in the area in which the application is taking place. Our experience at Union College and Williams College has been that there is a real advantage in having students enter the course knowing that virtually all the applications will focus on a single discipline—in this case, political science.

The level of presentation assumes no college–level mathematical or social science prerequisites. The philosophy underlying the approach we have taken in this book is based on the sense that we (mathematicians) have tended to make two errors in teaching nonscience students: We have overestimated their comfort with computational material, and we have underestimated their ability to handle conceptual material. Thus, while there is very little algebra (and certainly no calculus) in our presentation, we have included numerous logical arguments that students in the humanities and the social sciences will find accessible, but not trivial.

There are several ways in which the second edition differs from the first, most notably in the addition of a second author, for which the first author is extremely grateful. Those who used the answer book to the first edition will recognize Allison Pacelli as its author. There are also several structural ways in which the second edition differs from the first.

The first edition contained five main topics: escalation, conflict, yes-no voting, political power, and social choice. The first part of the text was made up of a single chapter devoted to each topic (Chapters 1–5), while the second part of the text revisited each topic (Chapters 6–10). For the second edition, we have completely reorganized the ten chapters, both in the order in which they appear and the choice of what material belongs in the first half of the book versus what belongs in the second half. We have also added some material to the existing chapters, and included two additional chapters devoted to a new topic: Fairness.

Thus, the second edition contains six main topics: social choice, yes-no voting, political power, conflict, fairness, and escalation. They are covered in this order (Chapters 1–6) in the first part of the text, and they are revisited in the second part of the text (Chapters 7–12). Within any given chapter, there is little reliance on material from earlier chapters, except for those devoted to the same topic. In addition to adding the two new chapters on fairness, we have introduced a new section to two of the topics in the first edition of the book, and updated material in a number of places. We also increased the number of exercises by roughly fifty percent, and corrected a couple of glitches brought to our attention by readers of the first edition.

The exercises are a crucial component of the book. They are not set up, however, to be used in the "daily homework" fashion that is typical in courses such as calculus, although such use is not ruled out. Rather, we have had more success assigning groups of problems to be done in a specified time period of one to two weeks, often discussing the problems at great length in class during the week or two that students are working on them. Another possibility is to have students work together on the problems in groups of two to four people.

As a final note, let us mention the obvious. Institutional resources permitting, a team–taught version of this course by a mathematician and a political scientist is extremely interesting.

Acknowledgements (Alan Taylor; from the First Edition)

The preparation of this book and the development of some of the material in it have been partially supported by the Alfred P. Sloan Foundation and the National Science Foundation. I am grateful to both.

There are two people to whom I owe an enormous debt: Steven J. Brams, from the Department of Politics at New York University, and William S. Zwicker, a colleague of mine in the Mathematics Department at Union. My debt to the work of Steve Brams will be obvious to anyone perusing this book; my debt to his inspiration is less obvious but just as real. Bill Zwicker has been equally influential. Chapters 3 and 8 (now 2 and 8) are directly taken from the joint work he and I have done over the past few years, but there is little in the text that he has not directly or indirectly influenced. It has been an honor and a pleasure to work with him.

Others have contributed to the text as well. Julius Barbanel and Karl Zimmermann taught courses using a manuscript version of this text and made many valuable suggestions. A number of students caught errors and typos, and a group of four Union students (Devra Eskin, Jennifer Johnson, Thomas Powers, and Melanie Rinaldi) suggested several of the exercises now found in Chapters 5 and 10 (now 1 and 6) as part of an Undergraduate Research Project supported by the National Science Foundation and directed by Bill Zwicker and myself.

I am grateful to John Ewing, Edward Packel, Thomas Quint, Stan Wagon, and several anonymous reviewers for a number of suggestions that significantly improved the text, and to Jerry Lyons and Liesl Gibson (and, for the second edition, to Achi Dosanjh) from the editorial staff at Springer-Verlag for their contributions.

Last, but not least, I thank my wife, Carolyn.

Acknowledgements (Allison Pacelli)

First, I would like to express my gratitude to Alan Taylor, not only for inviting me to co-author the second edition of this book, but also for introducing me to the field as an undergraduate. He has been a wonderful mentor and friend from my very first day of college when he asked me how would I divide a candy bar fairly among three people.

I would also like to thank Williams College for supporting the development of some of the new material in the book and my students Shomik Dutta, Zack Ulman, and Noam Yuchtman for suggesting the example of adjusted winner in the Middle East. I thank Ed Burger, a colleague of mine at Williams, for his encouragement and inspiration. Finally, I am forever grateful to my mother for her constant support and for playing all those math and logic games with me as a child.

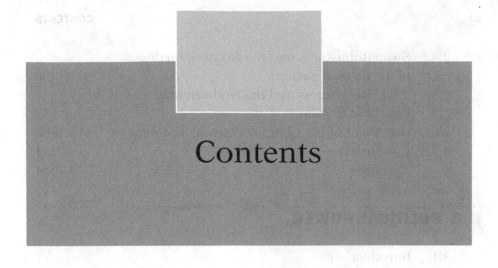

Contents

■ 3 POLITICAL POWER

■ 4 CONFLICT

■ 5 FAIRNESS

■ 6 ESCALATION

■ 7 MORE SOCIAL CHOICE

■ 8 MORE YES–NO VOTING

■ 9 MORE POLITICAL POWER

■ 10 MORE CONFLICT

■ 11 MORE FAIRNESS

■ 12 MORE ESCALATION

CHAPTER

1

Social Choice

■ 1.1 INTRODUCTION

In the present chapter we consider the situation wherein a group of voters is collectively trying to choose among several alternatives. When people speak of the area of "social choice," it is typically this context that they have in mind.

In the case where there are only two alternatives, the standard democratic process is to let each person vote for his or her preferred alternative, with the social choice (the "winner") being the alternative receiving the most votes. The situation, however, becomes complicated if there are more than two alternatives. In particular, if we proceed exactly as we did above where we had two alternatives, then we are not taking advantage of some individual comparisons among the several alternatives that could be made.

As a simple example of the kind of complication caused by more than two alternatives, consider the 1980 U.S. Senate race in New York among Alphonse D'Amato (a conservative), Elizabeth Holtzman (a liberal), and Jacob Javits (also a liberal). While we don't have complete information on the "preference orders" of the voters in New York at that time, we can make some reasonable estimates based on exit polls (showing, for example, that Javits's supporters favored Holtzman over D'Amato by a two-to-one margin). At any rate, for the sake of this example, we'll assume that each of the six possible (strict) preference orderings was held by the percentage of voters indicated below.

22%	23%	15%	29%	7%	4%
D	D	H	H	J	J
H	J	D	J	H	D
J	H	J	D	D	H

In addition to reflecting the intuitions confirmed by the exit polls, the above figures yield results coinciding with the known vote tallies of 45% for D'Amato, 44% for Holtzman, and 11% for Javits. The figures in the last two columns reflect the results of the actual exit poll that took place as described above. The middle two columns reflect a similar assumption as to the preference of Javits over D'Amato among Holtzman supporters, although the two-to-one ratio we use was not, to our knowledge, verified by exit polls. The first two columns are based (with no real justification) on the assumption that D'Amato supporters would be roughly evenly split between the two liberal candidates.

With each person voting for his or her top choice, D'Amato emerges as a (close) winner. On the other hand—and this is what is striking—notice that Holtzman would have defeated Javits in a two-person contest 66% to 34%, and she would have defeated D'Amato 51% to 49%. Thus, if we make use of all the information provided by the individual preference rankings, we get conflicting intuitions as to which alternative should reasonably be regarded as the "social choice."

It is precisely this kind of situation that motivates the considerations of the present chapter. The general framework will be as follows. There will be a set A whose elements will be called *alternatives* (or candidates) and typically denoted by a, b, c, etc. There will also be a set P whose elements will be called *people* (or voters) and typically denoted by p_1, p_2, p_3, etc. We shall assume that each person p in P has arranged the alternatives in a list (with no ties) according to preference. As above, we will picture these lists as being vertical with the alternatives displayed from most preferred on top to least preferred on the bottom. Such a list will be called an *individual preference list*, or, for brevity, a *ballot*. A sequence of ballots is called a *profile*. Our concern in this situation will be with so-called social choice procedures, where a social choice procedure is, intuitively, a fixed "recipe" for choosing an alternative based on the preference orderings of the individuals.

The mathematical notion underlying the concepts to be treated here is a simple one, but one of enormous importance in mathematics. This notion is that of a function. The definition runs as follows:

> **DEFINITION.** Suppose that X and Y are (not necessarily distinct) sets. Then a *function* from X to Y is a procedure that accepts each member of the set X as input and produces, for each such input, a single corresponding output that is a member of the set Y. The set X is called the domain of the function, and we speak of the procedure as being a function on the set X.

A "social choice procedure" is a special kind of function where a typical input is a profile and an output is a single alternative, or a single set of alternatives if we allow ties, or "NW" indicating that there is no winner.

Because of the importance of this notion, we record it here formally as a definition.

> **DEFINITION.** A *social choice procedure* is a function for which a typical input is a sequence of lists (without ties) of some set A (the set of alternatives) and the corresponding output is either an element of A, a subset of A, or "NW."

When discussing social choice procedures, we refer to the output as the "social choice" or "winner" if there is no tie, or the "social choice set" or "those tied for winner" if there is a tie.

In **Section 1.2** we begin with the case of two alternatives and a very elegant result (May's theorem) characterizing majority rule. In **Section 1.3** we will present six examples of social choice procedures. The examples are chosen to represent not only viable alternatives for real-world applications (e.g., the Hare system and the Borda count), but at least one extreme position (dictatorship) that will resurface later in two important theoretical contexts (Arrow's impossibility theorem and the Gibbard-Satterthwaite theorem). **Section 1.4**, on the other hand, introduces five apparently desirable properties (including independence of irrelevant alternatives and the Condorcet winner criterion, which are referred to below) that a given social choice procedure may or may not have. **Sections 1.5** and **1.6** then consider the obvious question: Which of the six social choice procedures satisfy which of the five desirable properties? Positive results are presented in **Section 1.5** while negative results are in **Section 1.6**.

In **Section 1.7** we foreshadow one of the cornerstones of social choice theory—Arrow's Theorem—that will be presented in Chapter 7 with the following impossibility theorem: There is no social choice procedure that satisfies independence of irrelevant alternatives, the Condorcet winner criterion, and always produces a winner. In **Section 1.8** we briefly discuss approval voting.

■ 1.2 MAY'S THEOREM FOR TWO ALTERNATIVES

In this section, we consider social choice procedures in which there are only two alternatives. The most common example of this is an election in which there are two candidates. If one alternative is represented by the letter "*a*" and the other by the letter "*b*," then there are only two possible preference lists (or ballots): the one that has *a* over *b* and the one that has *b* over *a*. We can think of the former preference list as a vote for alternative *a* and the latter as a vote for alternative *b*.

Most people would agree that there is really only one social choice procedure that suggests itself in this two-alternative situation: See which of and *a* and *b* has the most votes and declare it to be the winner. Indeed, this social choice procedure—typically called majority rule and formalized in the following definition—seems to be the cornerstone of our idea of democracy.

> **DEFINITION.** *Majority rule* is the social choice procedure for two alternatives in which an alternative is a winner if it appears at the top of at least half of the individual preference lists (equivalently, if at least half of the voters vote for that alternative).

In terms of a mathematical analysis of this two-alternative situation, there are two natural questions that suggest themselves: What properties of majority rule make it a compelling choice for democratic decision-making? Are there other social choice procedures in the two-alternative case that also satisfy these same desirable properties? Both of these questions are answered by the following elegant theorem of Kenneth May.

THEOREM. *(May, 1952) If the number of people is odd and each election produces a unique winner, then majority rule is the only social choice procedure for two alternatives that satisfies the following three conditions:*

1. *It treats all the voters the same: If any two voters exchange ballots, the outcome of the election is unaffected.*

2. *It treats both alternatives the same: If every voter reverses his or her vote (changing a vote for a to a vote for b and vice-versa), then the election outcome is reversed as well.*

3. *It is monotone: If some voter were to change his or her ballot from a vote for the loser to a vote for the winner, then the election outcome would be unchanged.*

We do not give a proof of May's theorem here, but a more general version is proved in Chapter 7. Condition (1) in May's theorem is called *anonymity* and condition (2) is called *neutrality*. In many ways, May's theorem tells us that for two alternatives, the search for a perfect voting system is really quite easy. Alas, things change dramatically when we move to the case of three or more alternatives, as we will now see.

■ 1.3 SIX EXAMPLES OF SOCIAL CHOICE PROCEDURES

We describe in this section six examples of social choice procedures. We have tried to pick a variety that includes some that are well known and often used, some that are inherently interesting, some that illustrate the desirable properties of the next section, and a final one (dictatorship) to illustrate that such procedures need not correspond to democratic choices (although many of the properties introduced in the next section will arise as attempts to isolate exactly such democratic choices). The examples are as follows.

Social Choice Procedure 1: Condorcet's Method

The social choice procedure known as Condorcet's method tries to take advantage of the success enjoyed by majority rule when there are only two alternatives. It does this by seeking an alternative that would, on the basis of the individual preference lists, defeat (or tie) every other alternative if the election had been between these two alternatives. Thus, with Condorcet's method, an alternative x is among the winners if for every other alternative y, at least half the voters rank x over y on their ballots. Although this method is typically attributed to the Marquis de Condorcet (1743–1794), it dates back at least to Ramon Llull in the thirteenth century.

To illustrate this idea of one-on-one competitions, suppose the preference lists are:

$$c \quad b \quad b \quad a \quad c$$
$$b \quad a \quad c \quad b \quad a$$
$$a \quad c \quad a \quad c \quad b$$

Then b defeats a in a one-on-one contest by a score of 3 to 2, since the first three voters rank b over a, while the last two voters rank a over b. The reader can also check that b defeats c by a score of 3 to 2, and that c defeats a by a score of 3 to 2.

Because b defeats each of the other alternatives in a one-on-one contest, it is the (unique) winner for this profile when Condorcet's method is used.

Social Choice Procedure 2: Plurality Voting

Plurality voting is the social choice procedure that most directly gener-alizes the idea of simple majority vote from the easy case of two alter-natives to the complicated case of three or more alternatives. The idea is simply to declare as the social choice(s) the alternative(s) with the largest number of first-place rankings in the individual preference lists.

Social Choice Procedure 3: The Borda Count

First popularized by Jean-Charles de Borda in 1781, the social choice procedure known as the Borda count takes advantage of the information regarding individual intensity of preference provided by looking at how high up in the preference list of an individual a given alternative occurs. More precisely, one uses each preference list to award "points" to each of n alternatives as follows: the alternative at the bottom of the list gets zero points, the alternative at the next to the bottom spot gets one point, the next one up gets two points and so on up to the top alternative on the list which gets $n - 1$ points. For each alternative, we simply add the points awarded it from each of the individual preference lists. The alternative(s) with the highest "Borda score" is declared to be the social choice.

Social Choice Procedure 4: The Hare System

The social choice procedure known as the Hare procedure was introduced by Thomas Hare in 1861, and is also known by names such as the "single transferable vote system" or "instant runoff voting." In 1862, John Stuart Mill spoke of it as being "among the greatest improvements yet made in the theory and practice of government." Today, it is used to elect public officials in Australia, Malta, the Republic of Ireland, and Northern Ireland.

The Hare system is based on the idea of arriving at a social choice by successive deletions of less desirable alternatives. More precisely, the procedure is as follows. We begin by deleting the alternative or alternatives occurring on top of the fewest lists. At this stage we have lists that are at least one alternative shorter than that with which we started. Now, we simply repeat this process of deleting the least desirable alternative or alternatives (as measured by the number of lists on top of which it, or they, appear). The alternative(s) deleted last is declared the winner.

Notice that if, at any stage, some alternative occurs at the top of more than half the lists, then that alternative will turn out to be the unique winner. However, an alternative occurring at the top of exactly half the lists—even if it is the only one doing so—is not necessarily the unique winner (although it must be among the winners).

Social Choice Procedure 5: Sequential Pairwise Voting with a Fixed Agenda

One typically thinks of an agenda as the collection of things to be discussed or decided upon. In the context of social choice theory, however, the term *agenda* refers to the *order* in which a fixed set of things will be discussed or decided upon. Thus, when we speak of a "fixed agenda," we are assuming we have a specified ordering $a, b, c, \ldots$ of the alternatives. (This ordering should not be confused with any of the individual preference orderings.) Sequential pairwise voting can be thought of as a series of one-on-one competitions among the alternatives as in Condorcet's method.

The procedure known as sequential pairwise voting with a fixed agenda runs as follows. We have a fixed ordering of the alternatives $a, b, c, \ldots$ called the agenda. The first alternative in the ordering is pitted against the second in a one-on-one contest. The winning alternative (or both, if there is a tie) is then pitted against the third alternative in the list in a one-on-one contest. An alternative is deleted at the end of any round in which it loses a one-on-one contest. The process is continued along the agenda until the "survivors" have finally met the last alternative in the agenda. Those remaining at the end are declared to be the social choices.

Social Choice Procedure 6: A Dictatorship

Of the six examples of social choice procedures we'll have at hand, this is the easiest to describe. Choose one of the "people" p and call this person the dictator. The procedure now runs as follows. Given the sequence of individual preference lists, we simply ignore all the lists except that of the dictator p. The alternative on top of p's list is now declared to be the social choice.

We shall illustrate the six social choice procedures with a single example that is somewhat enlightening in its own right.

Example:

Suppose we have five alternatives a, b, c, d, and e, and seven people who have individual preference lists as follows:

a	a	a	c	c	b	e
b	d	d	b	d	c	c
c	b	b	d	b	d	d
d	e	e	e	a	a	b
e	c	c	a	e	e	a

For each of our six procedures, we shall calculate what the resulting social choice is.

Condorcet's method: If we look at a one-on-one contest between alternatives a and b, we see that a occurs over b on the first three ballots and b occurs over a on the last four ballots. Thus, alternative b would defeat alternative a by a vote of 4 to 3 if they were pitted against each other. Similarly, alternative b would defeat alternative c (4 to 3, again) and alternative e (6 to 1). Thus, we have so far determined that neither a nor c nor e is the winner with Condorcet's method. But alternative d would defeat alternative b by a score of 4 to 3, and so b is not a winner either. This leaves only alternative d as a possibility for a winner. But alternative c handily defeats alternative d (5 to 2) and so d is also a non-winner. Hence, there is no winner with Condorcet's method.

Plurality: Since a occurs at the top of the most lists (three), it is the social choice when the plurality method is used.

Borda count: One way to find the Borda winner is to actually make a vertical column of values 4, 3, 2, 1, 0 to the left of the preference rankings. (Another way is to count the number of symbols occurring below the alternative whose Borda score is being calculated.) For example, alternative a receives a total of 14 points in the Borda system: four each for being in first place on the first three lists, none for being in last place on the fourth list and the seventh list, and one each for being in next to last place on the fifth and sixth lists. (Or, scanning the columns from left to right, we see that the number of symbols below a is $4 + 4 + 4 + 0 + 1 + 1 + 0$.) Similar calculations, again left for the reader, show that b gets 17 points, c and d each gets 16 points, and e gets only 7 points. Thus, the social choice is b when the Borda count is used.

Hare system: We decide which alternative occurs on the top of the fewest lists and delete it from all the lists. Since d is the only alternative not occurring at the top of any list, it is deleted from each list leaving the following:

$$
\begin{array}{ccccccc}
a & a & a & c & c & b & e \\
b & b & b & b & b & c & c \\
c & e & e & e & a & a & b \\
e & c & c & a & e & e & a
\end{array}
$$

Here, b and e are tied, each appearing on top of a single list, and so we now delete both of these from each list leaving the following:

$$a \quad a \quad a \quad c \quad c \quad c \quad c$$
$$c \quad c \quad c \quad a \quad a \quad a \quad a$$

Now, a occurs on top of only three of the seven lists, and thus is eliminated. Hence, c is the social choice when the Hare system is used.

Sequential pairwise voting with a fixed agenda $a\,b\,c\,d\,e$: We begin by pairing a against b in a one-on-one contest. Since b occurs higher up than a on a total of four of the seven lists (the last four), a is eliminated (having lost this one-on-one contest to b by a score of 4 to 3). Now, b goes against c and again emerges victorious by a score of 4 to 3, and so c is eliminated. Alternative b now takes on d, but winds up losing this one-on-one by a score of 4 to 3. Thus, b is eliminated and the final round pits d against e, which the reader can check is an easy win for d. Thus, d emerges as the social choice under sequential pairwise voting with this particular fixed agenda.

A dictatorship: We could pick any one of the seven people to be the dictator, but let's make it person number seven. Then the social choice is simply the alternative on top of the last list, which is e in this case.

Thus, our six examples of social choice procedures yield six different results when confronted by these particular preference lists. This raises the question of whether some procedures might be strictly better than others. But better in what ways? This we investigate in the next section.

■ 1.4 FIVE DESIRABLE PROPERTIES OF SOCIAL CHOICE PROCEDURES

The phrase *social choice* suggests that we are primarily interested in procedures that will select alternatives in a way that reflects, in some sense, the will of the people. A meaningful comparison of different procedures will require our having at hand some properties that are, at least intuitively, desirable. The social choice theory literature is not at all lacking in this regard. We shall, however, limit ourselves to the introduction of five such properties; more are introduced in the exercises. It should be noted that our choice of which properties to consider has been influenced, at least in part, by a desire to provide familiarity with some of the important ideas underlying major theorems in Chapter 7. In particular, a version of the property called "independence of irrelevant alternatives" will play a key role in Arrow's impossibility theorem.

The five properties are the following.

The Always-A-Winner Condition (AAW)

A social choice procedure is said to satisfy the *always-a-winner condition (AAW)* if every sequence of individual preference lists produces at least one winner.

The Condorcet Winner Criterion

An alternative x is said to be a *Condorcet winner* if it is the unique winner when Condorcet's method is used. Thus, x is a Condorcet winner provided that for every other alternative y, one finds x occurring above y on strictly more than half the lists. This defines what we mean by a Condorcet winner. For the definition of the "Condorcet winner criterion," we have the following:

A social choice procedure is said to satisfy the *Condorcet winner criterion* (CWC) provided that—if there is a Condorcet winner—then it alone is the social choice.

A sequence of preference lists often will not have a Condorcet winner, as we saw in the example in the last section. For those sequences of preference lists that do have a Condorcet winner, it always turns out to be unique; the Condorcet winner criterion is saying that, in this case, the unique Condorcet winner should be the unique winner produced by the social choice procedure. We should also point out that there are weaker versions of the Condorcet winner criterion that have been considered in the literature; see Fishburn (1973) or Nurmi (1987).

The Pareto Condition

A social choice procedure is said to satisfy the *Pareto condition* (or sometimes, for brevity, just *Pareto*) if the following holds for every pair x and y of alternatives:

If everyone prefers x to y, then y is not a social choice.

If we were not allowing ties, we could have said "the" social choice instead of "a" social choice in the statement of the Pareto condition (named after economist Vilfredo Pareto, who lived during the early part of the twentieth century). With ties, however, what we are saying is that if everyone finds x strictly preferable to y (recall that we are not allowing ties in the individual preference lists), then alternative y should not be the social choice and should not even be among the social choices if there is a tie.

Monotonicity

A social choice procedure is said to be *monotone* (or *monotonic*) provided that the following holds for every alternative x:

If x is the social choice (or tied for such) and someone changes his or her preference list by moving x up one spot (that is, exchanging x's position with that of the alternative immediately above x on his or her list), then x should still be the social choice (or tied for such).

The intuition behind the monotonicity condition is that if x is the social choice and someone changes his or her list in a way that is favorable to x (but not favorable to any other alternative) then x should remain the social choice. Monotonicity has also been called "non-perversity" in the literature. Indeed, a social choice procedure that is not monotone might well be regarded as perverse.

Independence of Irrelevant Alternatives

A social choice procedure is said to satisfy the condition of *independence of irrelevant alternatives* (IIA) provided that the following holds for every pair of alternatives x and y:

If the social choice set includes x but not y, and one or more voters change their preferences, but no one changes his or her mind about whether x is preferred to y or y to x, then the social choice set should not change so as to include y.

The point here is that if a preference list is changed but the relative positions of x and y to each other are not changed, then the new list can be described as arising from upward and downward shifts of alternatives other than x and y. Changing preferences toward these other alternatives should, intuitively, be irrelevant to the question of social preference of x to y or y to x.

Of course, if we start with x a winner and y a nonwinner, and people move some other alternative z around, then we cannot hope to conclude that x is still a winner. After all, everyone may have moved z to the top of their list. Independence of irrelevant alternatives is simply saying that y should remain a nonwinner.

To feel comfortable with these properties, one needs to see some specific social choice procedures that provide illustrations of the properties themselves and—perhaps more importantly—examples of their failure. This occurs in the next two sections.

■ 1.5 POSITIVE RESULTS—PROOFS

From the previous two sections we have at hand six social choice procedures (Condorcet's method, purality, Borda, Hare, sequential pairwise, and dictatorship) and five properties (always a winner, the Condorcet winner criterion, Pareto, monotonicity, and independence of irrelevant alternatives) pertaining to such procedures. Which procedures satisfy which properties? The answer is given in the following table (where a "yes" indicates the property holds for the given procedure).

	AAW	CWC	Pareto	Mono	IIA
Condorcet		Yes	Yes	Yes	Yes
Plurality	Yes		Yes	Yes	
Borda	Yes		Yes	Yes	
Hare	Yes		Yes		
Seq Pairs	Yes	Yes		Yes	
Dictator	Yes		Yes	Yes	Yes

Our goal in this section is to prove the nineteen positive results in the chart. The first five positive results—that all but Condorcet's method always produce at least one winner—are collected together in Proposition 1 and treated in a somewhat dismissive manner. Each of the other results will be stated as a proposition and provided with a complete proof that emphasizes the structural aspects of the definitions of the properties. That is, we clearly indicate that we are dealing with an arbitrary sequence of preference lists, that we are making explicit assumptions, and that there is a specific thing to be shown in each of the proofs. This, however, tends to obscure the heart of the arguments and so we have added, for each, an informal brief explanation of why the proposition is true. Probably little, if anything, is lost by simply regarding these brief explanations as the formal proofs.

PROPOSITION 1. *The plurality procedure, the Borda count, the Hare system, sequential pairwise voting with a fixed agenda, and a dictatorship all satisfy the always-a-winner condition.*

PROOF. For each of the procedures, the description makes it clear that there is at least one winner for every profile.

PROPOSITION 2. *Condorcet's method satisfies the Condorcet winner criterion.*

PROOF. Assume the social choice procedure being used is Condorcet's method and that we have an arbitrary sequence of individual preference lists where there is an alternative x that is a Condorcet winner. Then, by definition of Condorcet winner, x is the unique winner when Condorcet's method is used.

Briefly, by definition, a Condorcet winner is the unique winner when Condorcet's method is used.

PROPOSITION 3. *Sequential pairwise voting with a fixed agenda satisfies the Condorcet winner criterion.*

PROOF. Assume the social choice procedure being used is sequential pairwise voting with a fixed agenda and assume that we have an arbitrary sequence of preference lists where there is an alternative x that is the Condorcet winner. We want to show that x is the social choice. In sequential pairwise voting the social choices are the alternatives that are not eliminated at any stage in the sequence of one-on-one contests. But being a Condorcet winner means that precisely this kind of one-on-one contest is always won. Thus, x is the (only) social choice. This completes the proof.

Briefly, a Condorcet winner always wins the kind of one-on-one contest that is used to produce the social choice in sequential pairwise voting.

PROPOSITION 4. Condorcet's method satisfies the Pareto condition.

PROOF. Assume the social choice procedure being used is Condorcet's method and that we have an arbitrary sequence of individual preference lists where everyone prefers alternative x to alternative y. Then x defeats y in a one-on-one contest and so y cannot be a winner with Condorcet's method.

Briefly, if everyone prefers x to y, then y fails to defeat x in a one-on-one contest, and so y cannot be a winner with Çondorcet's method.

PROPOSITION 5. The plurality procedure satisfies the Pareto condition.

PROOF. Assume the social choice procedure being used is the plurality procedure and assume that we have an arbitrary sequence of preference lists where everyone prefers alternative x to alternative y. We must show that y is not a social choice. But this is easy, since the social choice is the alternative on top of the most lists, and y can't be on top of any list, since x occurs higher up than y on every list. This completes the proof.

Briefly, if everyone prefers x to y, then y is not on top of any list (let alone a plurality) and thus y is certainly not a social choice.

PROPOSITION 6. The Borda count satisfies the Pareto condition.

PROOF. Assume the social choice procedure being used is the Borda count and assume that we have an arbitrary sequence of preference lists where everyone prefers alternative x to alternative y. We must show that y is not a social choice. Since x occurs higher than y on each of the preference lists, x receives more points from each list than does y. Thus, when we add up the points awarded from each list we clearly have a strictly higher total for x than for y. This does not guarantee that x is the social choice, but it certainly guarantees that y is not, and this is what we wanted to show. This completes the proof.

Briefly, if everyone prefers x to y, then x receives more points from each list than y. Thus, x receives a higher total than y and so y is certainly not a social choice.

PROPOSITION 7. *The Hare system satisfies the Pareto condition.*

PROOF. Assume the social choice procedure being used is the Hare system and assume that we have an arbitrary sequence of preference lists where everyone prefers alternative x to alternative y. We must show that y is not a social choice. Notice again that y is not on top of any list. Thus, y is among the alternatives immediately deleted, since it occurs at the top of no lists and that is as few as you can get. This shows that y is not a social choice and completes the proof.

Briefly, if everyone prefers x to y, then y is not on top of any list. Thus, y is eliminated at the very first stage. Hence, y is not a social choice.

PROPOSITION 8. *The dictatorship procedure satisfies the Pareto condition.*

PROOF. Assume the social choice procedure being used is a dictatorship and assume that we have an arbitrary sequence of preference lists where everyone prefers alternative x to alternative y. We must show that y is not a social choice. But if everyone prefers x to y then, in particular, the dictator does and so y is not on top of the dictator's list. Since the social choice is whichever alternative happens to be on top of the dictator's list, this shows that y is not a social choice and completes the proof.

Briefly, if everyone prefers x to y, then, in particular, the dictator does. Hence, y is not on top of the dictator's list and so is not a social choice.

PROPOSITION 9. *Condorcet's method satisfies monotonicity.*

PROOF. Assume the social choice procedure being used is Condorcet's method and that we have an arbitrary sequence of individual

preference lists yielding *x* as a social choice. Now assume that someone exchanges *x*'s position with that of the alternative directly above *x*
on his or her list. We want to show that *x* is still a social choice. But
the change in the single list described above affects only the one-on-
one contest between *x* and the alternative with which it was switched.
Clearly *x* not only still wins this contest, but by a larger margin. Thus, *x*
is still a social choice using Condorcet's method.

Briefly, moving *x* up on some list only improves *x*'s chances in one-
on-one contests.

PROPOSITION 10. *The plurality procedure satisfies monotonicity.*

PROOF. Assume the social choice procedure being used is the plurality
procedure and assume that we have an arbitrary sequence of preference
lists yielding *x* as a social choice. Now assume that someone exchanges
x's position with that of the alternative above *x* on his or her list. We
want to show that *x* is still a social choice. But since *x* was originally a
social choice, *x* was at least tied for being on top of the most lists. The
change in the single list described above neither decreases the number
of lists that *x* is on top of nor increases the number of lists that any
other alternative is on top of. Thus, *x* is still among the social choices
and so the proof is complete.

Briefly, if *x* is on top of the most lists (or tied for such), then moving
x up one spot on some list (and making no other changes) certainly
preserves this.

PROPOSITION 11. *The Borda count satisfies monotonicity.*

PROOF. Assume the social choice procedure being used is the Borda
count and assume that we have an arbitrary sequence of preference lists
yielding *x* as a social choice. Now assume that someone exchanges *x*'s
position with that of the alternative above *x* on his or her list. We want
to show that *x* is still a social choice. But the change in the single list
described above simply adds one point to *x*'s total, subtracts one point
from that of the other alternative involved, and leaves the scores of all
the other alternatives unchanged. Thus, *x* is still a social choice and so
the proof is complete.

Briefly, swapping x's position with the alternative above x on some list adds one point to x's score and subtracts one point from that of the other alternative; the scores of all other alternatives remain the same.

PROPOSITION 12. *Sequential pairwise voting with a fixed agenda satisfies monotonicity.*

PROOF. Assume the social choice procedure being used is sequential pairwise voting with a fixed agenda and assume that we have an arbitrary sequence of preference lists yielding x as a social choice. Now assume that someone exchanges x's position with that of the alternative above x on his or her list. We want to show that x is still a social choice. But the change in the single list described above affects only the one-on-one contest between x and the alternative with which it was switched. Clearly x not only still wins this contest, but by a larger margin. Thus, x is still a social choice and so the proof is complete.

Briefly, moving x up on some list only improves x's chances in one-on-one contests.

PROPOSITION 13. *A dictatorship satisfies monotonicity.*

PROOF. Assume the social choice procedure being used is a dictatorship and assume that we have an arbitrary sequence of preference lists yielding x as a social choice. Now assume that someone exchanges x's position with that of the alternative above x on his or her list. We want to show that x is still a social choice. Since x is a social choice, we know that x is on top of the dictator's list. Thus, the exchange described above could not have taken place in the dictator's list since there is no alternative above x with which to exchange it. Thus, x is still on top of the dictator's list and so x is still the social choice. This completes the proof.

Briefly, if x is the social choice then x is on top of the dictator's list. Hence, the exchange of x with some alternative immediately above x must be taking place on some list other than that of the dictator. Thus, x is still the social choice.

PROPOSITION 14. *Condorcet's method satisfies independence of irrelevant alternatives.*

PROOF. Assume the social choice procedure being used is Condorcet's method and that we have an arbitrary sequence of individual preference lists yielding x as a winner and y as a non-winner. Thus x defeats every other alternative in a one-on-one contest. Now suppose that preference lists are changed but no one changes his or her mind about whether x is preferred to y or y to x. We want to show that y is not among the social choices, which simply means that y does not defeat every other alternative in a one-on-one contest. But because no one who had x over y changed this to y over x, we still have y losing to x in a one-on-one contest. Hence, y is not a social choice using Condorcet's method.

Briefly, if x is a Condorcet winner and thus defeats every other alternative one on one, and no one who had x over y moves y over x, then y still loses to x one on one, and so is not a winner with Condorcet's method.

PROPOSITION 15. *A dictatorship satisfies independence of irrelevant alternatives.*

PROOF. Assume the social choice procedure being used is a dictatorship and assume that we have an arbitrary sequence of preference lists yielding x as a winner and y as a nonwinner. Thus, x is on top of the dictator's list. Now suppose that preference lists are changed but no one changes his or her mind about whether x is preferred to y or y to x. We want to show that y is not now among the social choices, which simply means that y is not now on top of the dictator's list. But the dictator's list still has x over y (although x may no longer be on top). Thus, y is not on top of the dictator's list and so y is not a social choice. This completes the proof.

Briefly, if x is the social choice and no one—including the dictator—changes his or her mind about x's preference to y, then y cannot wind up on top of the dictator's list. Thus, y is not the social choice.

···

■ 1.6 NEGATIVE RESULTS—PROOFS

In the previous section we were concerned with the properties that held for the various social choice procedures under consideration. Our concern here, however, is with those that fail. In a sense, these results are somewhat more striking than those of the previous section, since most of the procedures and properties seem to be quite reasonable, and one certainly expects reasonable procedures to satisfy reasonable properties.

The following table indicates which properties fail for which procedures. It is simply the "dual" of the table occurring at the beginning of the last section.

	AAW	CWC	Pareto	Mono	IIA
Condorcet	No				
Plurality		No			No
Borda		No			No
Hare		No		No	No
Seq Pairs			No		No
Dictator		No			

Our goal in this section is to prove the eleven negative results in the chart. Again, each will be stated as a proposition. The structure of the proofs, however, will he quite different from those of the last section. In particular, the properties we are dealing with all assert that regardless of what sequence of preference lists we happen to be considering, some pathological thing does not take place. Thus, in proving that a property holds, as we did in the last section, we had to consider an arbitrary sequence of preference lists (as opposed to a particular sequence of our choosing). On the other hand, in proving that a property *fails*, we need only produce *one example* of a sequence of preference lists exhibiting the pathological behavior mentioned in the property.

PROPOSITION 1. *Condorcet's method fails to satisfy the always-a-winner condition.*

PROOF. We have already seen a proof of this; indeed, the sequence of individual preference lists in the example in **Section 1.3** had no winner with Condorcet's method. But this result is important enough to justify presenting the simplest example of this failure of Condorcet's method to produce a winner. This example involves only three alternatives and three voters. It is known as "Condorcet's voting paradox" or the "voting paradox of Condorcet."

Condorcet's Voting Paradox

Voter 1	Voter 2	Voter 3
a	b	c
b	c	a
c	a	b

Alternative a is not a winner, because it is defeated by alternative c (by a score of 2 to 1).

Alternative b is not a winner, because it is defeated by alternative a (by a score of 2 to 1).

Alternative c is not a winner, because it is defeated by alternative b (by a score of 2 to 1).

PROPOSITION 2. *The plurality procedure fails to satisfy the Condorcet winner criterion.*

PROOF. Consider the three alternatives a, b, and c and the following sequence of nine preference lists grouped into "blocs" of sizes four, three, and two.

Voters 1–4	Voters 5–7	Voters 8 and 9
a	b	c
b	c	b
c	a	a

With the plurality procedure, alternative a is clearly the social choice since it has four first-place votes to three for b and two for c. On the other hand, we claim that b is a Condorcet winner. That is, b would defeat a by a score of 5 to 4 in one-on-one competition, and b would defeat c by a score of 7 to 2 in one-on-one competition. Thus, the Condorcet winner b is not the social choice, and so the Condorcet winner criterion fails for the plurality procedure.

PROPOSITION 3. *The Borda count does not satisfy the Condorcet winner criterion.*

PROOF. Consider the three alternatives *a*, *b*, and *c* and the following sequence of five preference lists grouped into voting blocs of size three and two.

Voters 1–3	Voters 4 and 5
a	b
b	c
c	a

The Borda count produces *b* as the social choice since it gets a total of 7 points $(1+1+1+2+2)$ to 6 points for *a* $(2+2+2+0+0)$ and 2 points for *c* $(0+0+0+1+1)$. However, *a* is clearly the Condorcet winner, defeating each of the other alternatives by a score of 3 to 2 in one-on-one competitions. Since the Condorcet winner is not the social choice in this situation, we have that the Borda count does not satisfy the Condorcet winner criterion.

PROPOSITION 4. *The Hare procedure does not satisfy the Codorcet winner criterion.*

PROOF. Consider the five alternatives *a*, *b*, *c*, *d*, and *e* and the following sequence of seventeen preference lists grouped into blocs of size five, four, three, three, and two:

Voters 1–5	Voters 6–9	Voters 10–12	Voters 13–15	Voters 16 and 17
a	e	d	c	b
b	b	b	b	c
c	c	c	d	d
d	d	e	e	e
e	a	a	a	a

We claim first that *b* is the Condorcet winner. The results and scores are as follows: *b* defeats *a* (12 to 5), *b* defeats *c* (14 to 3), *b* defeats *d* (14 to 3), *b* defeats *e* (13 to 4). On the other hand, the social choice according to the Hare procedure is definitely not *b*; in the first stage of the procedure alternative *b* is deleted from all the lists since it has only two first-place votes. This much already shows that the Condorcet winner is not a social choice, and thus the proof is complete.

PROPOSITION 5. *A dictatorship does not satisfy the Condorcet winner criterion.*

PROOF. Consider the three alternatives *a*, *b*, and *c* and the following three preference lists:

Voter 1 Voter 2 Voter 3

a	c	c
b	b	b
c	a	a

Assume that Voter 1 is the dictator. Then *a* is the social choice, although *c* is clearly the Condorcet winner since it defeats both others by a score of 2 to 1.

PROPOSITION 6. *Sequential pairwise voting with a fixed agenda does not satisfy the Pareto condition.*

PROOF. Consider the four alternatives *a*, *b*, *c*, and *d* and suppose that this ordering of the alternatives is also the agenda. Consider the following sequence of three preference lists:

Voter 1   Voter 2   Voter 3

a	c	b
b	a	d
d	b	c
c	d	a

Clearly, everyone prefers *b* to *d*. But with the agenda *a b c d* we see that alternative a first defeats *b* by a score of 2 to 1, and then *a* loses to *c* by this same score. Alternative *c* now goes on to face *d*, but *d* defeats *c* again by a 2 to 1 score. Thus, alternative *d* is the social choice even though everyone prefers *b* to *d*. This shows that Pareto fails.

PROPOSITION 7. *The Hare procedure does not satisfy monotonicity.*

PROOF. Consider the alternatives $a, b,$ and c and the following sequence of seventeen preference lists grouped into voting blocs of size seven, five, four, and one.

Voters 1–7	Voters 8–12	Voters 13–16	Voter 17
a	c	b	b
b	a	c	a
c	b	a	c

We delete the alternatives with the fewest first place votes. In this case, that would be alternatives c and b with only five first place votes each as compared to seven for a. But now a is the only alternative left, and so it is the social choice when the Hare procedure is used.

Now suppose that the single voter on the far right changes his or her list by interchanging a with the alternative that is right above a on this list. This apparently favorable-to-a-change yields the following sequence of preference lists:

Voters 1–7	Voters 8–12	Voters 13–16	Voter 17
a	c	b	a
b	a	c	b
c	b	a	c

If we apply the Hare procedure again, we delete the alternative with the fewest first place votes. In this case, that alternative is b with only four. But the reader can now easily check that with b so eliminated, alternative c is on top of nine of the seventeen lists. Alternative a is deleted and so c is the social choice. This change in social choice from a to c shows that the Hare system does not satisfy monotonicity.

PROPOSITION 8. *The plurality procedure does not satisfy independence of irrelevant alternatives.*

PROOF. Consider the alternatives $a, b,$ and c and the following sequence of four preference lists:

Voter 1	Voter 2	Voter 3	Voter 4
a	a	b	c
b	b	c	b
c	c	a	a

Clearly, alternative a is the social choice when the plurality procedure is used. In particular, a is a winner and b is a nonwinner. Now suppose that Voter 4 changes his or her list by moving the alternative c down between b and a. The lists then become:

Voter 1	Voter 2	Voter 3	Voter 4
a	a	b	b
b	b	c	c
c	c	a	a

Notice that we still have b over a in Voter 4's list. However, plurality voting now has a and b tied for the win with two first place votes each. Thus, although no one changed his or her mind about whether a is preferred to b or b to a, the alternative b went from being a nonwinner to being a winner. This shows that independence of irrelevant alternatives fails for the plurality procedure.

PROPOSITION 9. *The Borda count does not satisfy independence of irrelevant alternatives.*

PROOF. Consider the alternatives $a, b,$ and c and the following sequence of five preference lists grouped into voting blocs of size three and two.

Voters 1–3	Voters 4 and 5
a	c
b	b
c	a

The Borda count yields a as the social choice since it gets 6 points $(2 + 2 + 2 + 0 + 0)$ to only five for b $(1 + 1 + 1 + 1 + 1)$ and four for c $(0 + 0 + 0 + 2 + 2)$. But now suppose that Voters 4 and 5 change their list by lowering c from first to second position, but still maintaining the same relative position of b over a. The lists then look as follows:

Voters 1–3   Voters 4 and 5

Voters 1–3	Voters 4 and 5
a	b
b	c
c	a

The Borda count now yields b as the social choice with seven points to only six for a and two for c. Thus, the social choice has changed from a to b although no one changed his or her mind about whether a is preferred to b or b to a. Hence, independence of irrelevant alternatives fails for the Borda count.

PROPOSITION 10. *The Hare procedure fails to satisfy independence of irrelevant alternatives.*

PROOF. Consider the alternatives a, b, and c and the same sequence of four preference lists that we used in Proposition 8:

Voter 1	Voter 2	Voter 3	Voter 4
a	a	b	c
b	b	c	b
c	c	a	a

Alternative a is the social choice when the Hare procedure is used because alternatives b and c have only one first-place vote each. In particular, a is a winner and b is a nonwinner. Now suppose, as we did in the proof of Proposition 8, that Voter 4 changes his or her list by moving the alternative c down between b and a. The lists then become:

Voter 1	Voter 2	Voter 3	Voter 4
a	a	b	b
b	b	c	c
c	c	a	a

Notice that we still have b over a in Voter 4's list. Under the Hare procedure, we now have a and b tied for the win, since each has half the first place votes. Thus, although no one changed his or her mind about whether *a* is preferred to *b* or *b* to *a*, the alternative *b* went from being a nonwinner to being a winner. This shows that independence of irrelevant alternatives fails for the Hare procedure.

PROPOSITION 11. *Sequential pairwise voting with a fixed agenda fails to satisfy independence of irrelevant alternatives.*

PROOF. Consider the alternatives *c*, *b*, and *a* and assume this reverse alphabetical ordering is the agenda. Consider the following sequence of three preference lists:

Voter 1	Voter 2	Voter 3
c	a	b
b	c	a
a	b	c

In sequential pairwise voting, *c* would defeat *b* by the score of 2 to 1 and then lose to *a* by this same score. Thus, *a* would be the social choice (and thus *a* is a winner and *b* is a nonwinner). But now suppose that Voter 1 moves *c* down between *b* and *a*, yielding the following lists:

Voter 1	Voter 2	Voter 3
b	a	b
c	c	a
a	b	c

Now, *b* first defeats *c* and then *b* goes on to defeat *a*. Hence, the new social choice is *b*. Thus, although no one changed his or her mind about whether *a* is preferred to *b* or *b* to *a*, the alternative *b* went from being a nonwinner to being a winner. This shows that independence of irrelevant alternatives fails for sequential pairwise voting with a fixed agenda.

This completes our task of verifying the eleven "no" entries from the chart of procedures and properties at the beginning of this section. One should, however, find the results of this section to be somewhat unsettling. The properties, after all, seem to be quite reasonable, as do most of the procedures. Why haven't we presented a number of natural procedures that satisfy *all* of these properties and more? We turn to this question next.

■ 1.7 A GLIMPSE OF IMPOSSIBILITY

In Chapter 7 we shall return to the issue of social choice and present the single most famous theorem in the field: Arrow's impossibility theorem. The natural context for Arrow's theorem, however, is slightly different from the context in which we have explored social choice in the present chapter. Nevertheless, this section previews the kind of difficulty that Arrow's theorem shows is unavoidable. We will do this by stating and proving an impossibility theorem in the context with which we have worked in the present chapter. The proof of this theorem, like that of Arrow's theorem, makes critical use of the voting paradox of Condorcet.

Recall that in **Section 1.4** we introduced five desirable properties of social choice procedures: the always-a-winner condition, the Pareto condition, the Condorcet winner criterion, monotonicity, and independence of irrelevant alternatives. Of the six social choice procedures we looked at, only Condorcet's method and a dictatorship satisfied independence of irrelevant alternatives, and only Condorcet's method and sequential pairwise voting satisfied the Condorcet winner criterion. None of the six procedures satisfied all five of the desirable properties.

Suppose we were to seek a social choice procedure that satisfies all five of our desirable properties. One possibility is to start with one of the six procedures that we looked at and to modify it in such a way that a property that was not satisfied by the original procedure would be satisfied by the new version. For example, there is a very natural way to modify a procedure so that the Condorcet winner criterion becomes satisfied: If there is a Condorcet winner, then it is the social choice; otherwise, apply the procedure at hand.

It is tempting to think that if we modify a dictatorship in the above way, then we will have a social choice procedure that satisfies all five of the desirable properties we discussed in this chapter. This turns out not to be the case (and we will say why in a moment). But maybe there are other ways to alter one or more of the procedures from this chapter so that the result will satisfy all the desirable properties. Or maybe there are procedures that look very different from the ones we presented in this chapter that already satisfy these desirable properties. Or maybe no such procedures have ever been found, but that one will be found a hundred years from now.

No way.

There is no social choice procedure that satisfies all five of the desirable properties that we listed in **Section 1.4**. We are not just saying that none of the six procedures we looked at satisfies all five of the desirable properties—we already know that. We are not just saying that these procedures can't be altered to yield one that satisfies all five of the desirable properties. We are not just saying that no one has yet found a social choice procedure that satisfies all five of the desirable properties. We are saying that no one will *ever* find a social choice procedure that satisfies these five desirable properties. In fact, more is true:

THEOREM. *There is no social choice procedure for three or more alternatives that satisfies the always-a-winner criterion, independence of irrelevant alternatives, and the Condorcet winner criterion.*

We will assume that we have a social choice procedure that satisfies both independence of irrelevant alternatives and the Condorcet winner criterion. We will then show that if this procedure is applied to the profile that constitutes Condorcet's voting paradox (**Section 1.6**), then it produces no winner. Because any procedure satisfying IIA and the CWC fails to satisfy AAW, it follows that no procedure can satisfy all three criteria.

PROOF. Assume that we have a social choice procedure that satisfies both independence of irrelevant alternatives and the Condorcet winner criterion. Consider the following profile from the voting paradox of Condorcet:

$$
\begin{array}{ccc}
a & c & b \\
b & a & c \\
c & b & a
\end{array}
\tag{1}
$$

CLAIM 1. The alternative a is a nonwinner.

PROOF. Consider the following profile (obtained by moving alternative *b* down in the third preference list from the voting paradox profile):

$$
\begin{array}{ccc}
a & c & c \\
b & a & b \\
c & b & a
\end{array}
\qquad (2)
$$

Notice that *c* is a Condorcet winner for profile (2) (defeating both other alternatives by a margin of 2 to 1). Thus, our social choice procedure (which we are assuming satisfies the Condorcet winner criterion) must produce *c* as the only winner. Thus, *c* is a winner and *a* is a nonwinner for this profile. (We are not done proving the claim because this is not the voting paradox profile.)

Suppose now that the third voter moves *b* up on his or her preference list. The profile then becomes that of the voting paradox (since we just undid what we did earlier). We want to show that *a* is still a nonwinner.

But no one changed his or her mind about whether *c* is preferred to *a* or *a* is preferred to *c*. Thus, because our procedure is assumed to satisfy independence of irrelevant alternatives, and because we had *c* as a winner and *a* as a nonwinner in the profile with which we began the proof of the claim, we can conclude that *a* is still a nonwinner when the procedure is applied to profile (1). This proves the claim.

CLAIM 2. The alternative *b* is a nonwinner.

PROOF. Consider the following profile (obtained by moving alternative *c* down in the second preference list from the voting paradox profile):

$$
\begin{array}{ccc}
a & a & b \\
b & c & c \\
c & b & a
\end{array}
\qquad (3)
$$

Notice that *a* is a Condorcet winner for profile (3) (defeating both other alternatives by a margin of 2 to 1). Thus, our social choice procedure (which we are assuming satisfies the Condorcet winner criterion) must produce *a* as the only winner. Thus, *a* is a winner and *b* is a nonwinner for this profile. (We are again not done proving the claim because this is not the voting paradox profile.)

Suppose now that the second voter moves c up on his or her preference list. The profile then becomes that of the voting paradox (since we just undid what we did earlier). We want to show that b is still a nonwinner.

But no one changed his or her mind about whether a is preferred to b or b is preferred to a. Thus, because our procedure is assumed to satisfy independence of irrelevant alternatives, and because we had a as a winner and b as a nonwinner in the profile with which we began the proof of the claim, we can conclude that b is still a nonwinner when the procedure is applied to profile (1). This proves the claim.

CLAIM 3. The alternative c is a nonwinner.

PROOF. We leave this for the reader (see Exercise 40).

The above three claims show that when our procedure is confronted with the voting paradox profile, it produces *no* winner. Thus, any social choice procedure satisfying IIA and the CWC fails to satisfy AAW. This completes the proof.

This is only part of the remarkable story of the difficulty with "reflecting the will of the people." More of the story will be told in Chapter 7.

··

■ 1.8 APPROVAL VOTING

The voting systems we have considered so far are social choice procedures: a collection of individual preference lists (without ties) is the input, and the output is a single or possibly a collection of alternatives. There are, however, other types of voting systems. Here we consider one of the most popular alternative methods—approval voting. Approval voting was explicitly proposed in the 1971 Ph.D. thesis of Robert Weber at Yale University. Since then, Steven Brams, a political scientist at NYU, and Peter Fishburn, a former researcher at Bell Laboratories, have done much more research on and promotion of approval voting. Under approval voting, given a set A of alternatives, each voter votes for (or "approves of") as many alternatives as he or she chooses. The voters do not rank the alternatives. The social choice is the alternative (or set of alternatives) with the largest number of votes.

For example, suppose that there are three alternatives and five voters. The ballots might look as follows, where each column consists of the set of all alternatives approved of by the corresponding voter. Remember that the ordering within the column is arbitrary (alphabetical in this case); no ranking of alternatives is indicated.

$$
\begin{array}{ccccc}
a & a & a & b & c \\
  & c & b &   & c \\
  &   & c &   &
\end{array}
$$

In this example, three voters approve of alternative a, two voters approve of alternative b, and four voters approve of alternative c; alternative c is therefore the social choice.

Many professional societies—including the American Mathematical Society, the Mathematical Association of America, and the National Academy of Sciences—use approval voting for some elections. Since 1996, approval voting has been used by the United Nations to elect the Secretary-General; it has also been used in government elections in Pennsylvania, Oregon, Eastern Europe, and the Soviet Union.

Supporters of approval voting argue that it is much easier to understand than some other procedures. It allows individual voters to equally value two or more alternatives unlike the social choice procedures we have looked at previously which do not allow ties. Because voters essentially need only say yes or no for each alternative, approval voting may be easier for the voters than other procedures which require the voters to rank each alternative. Opponents, however, argue that since approval voting does not use as much information about the voters' preferences, the resulting social choice does not as accurately reflect the will of the people.

Another major argument in support of approval voting is that it will reduce negative campaigning. Negative campaigning is more effective in a plurality system, since only first-place votes matter. If a candidate is not a voter's first choice, then it makes no difference whether that alternative is second or last in the voter's opinion. It doesn't matter, therefore, if negative campaigning further lowers a candidate's status in a voter's eyes. When voters can vote for more than one alternative though, candidates have a major incentive to remain well respected by as many voters as possible since a voter may decide to vote for his top two, three, four, or more candidates. It is quite possible that negative campaigning would therefore decrease with approval voting because negative campaigning is often looked down upon by voters. Note that this argument applies not only to approval voting, but to any system in which candidates can benefit by being high on (even if not on top of) a voter's preference list.

Another argument in support of approval voting is that it eliminates the effect of *spoiler candidates*, candidates who cannot feasibly win an election but sometimes alter the outcome of an election. For example, many Gore supporters in 2000 blamed Ralph Nader voters when George W. Bush was elected. Since there is reason to believe that the majority of Nader voters preferred Gore to Bush, those voters would likely have voted for Gore had Nader not been an alternative. It is possible then that the presence of Nader as a candidate caused Bush to win over Gore. Supporters of third-party candidates often face the difficult dilemma of voting for their true first-choice candidate, or strategically voting for their second-choice candidate since their first choice is unlikely to win. With approval voting, voters have the option of voting for both; they are able to express their support for their desired candidate while preventing that support from throwing the election to their least favorite candidate. Again, it is worth noting that the social choice procedures which use the voter's full ranking of the candidates also reduce the effect of spoiler candidates.

Approval voting allows more flexibility than plurality. Under approval voting, a voter still has the option of voting solely for their first-place alternative, but has the flexibility to vote for more. Opponents of approval voting argue though that this flexibility is a drawback; one can show that depending on where the voters draw the line between approval and disapproval, almost any candidate can win. For example, before an election using approval voting, the president of the Mathematical Association of America issued the following statement to the voters: "Suppose there are three candidates of whom two are outstanding. Suppose the third is a person you believe is not yet ready for office but whom you decide to vote for as a means of encouragement (in addition to voting for your favorite). If enough voters reason that way, you will elect that person now." (L. Gilman, FOCUS). While this may be alarming, it may be reasonable to assume that if the voters understand the system, this situation would not occur. One might argue that if a voter truly believes someone not ready for office, then he or she does not "approve" of and therefore should not vote for that candidate.

■ 1.9 CONCLUSIONS

We began the chapter by looking at the 1980 U.S. Senate race in New York where Alphonse D'Amato defeated Elizabeth Holtzman and Jacob Javits, even though reasonable assumptions suggest that Holtzman could have beaten either D'Amato or Javits in a one-on-one contest. (In terminology from later in the chapter, Holtzman was a Condorcet winner.) This introduction was meant to suggest some potential difficulties in producing a "reasonable" social choice when there are three or more alternatives. In terms of mathematical preliminaries, we introduced the notion of a function and defined a social choice procedure to be a special kind of function where a typical input is a sequence of preference lists and the corresponding output is either a single alternative (the social choice), a collection of alternatives, or the symbol NW indicating no winner.

The chapter introduced six social choice procedures—Condorcet's method, plurality voting, the Borda count, the Hare system, sequential pairwise voting with a fixed agenda, and a dictatorship—and five apparently desirable properties that pertain to such procedures—the always-a-winner condition, the Pareto condition, the Condorcet winner criterion, monotonicity, and independence of irrelevant alternatives. Asking the thirty obvious questions about which procedures satisfy which properties produced both affirmative answers and negative answers. Among the negative answers were some striking results: the Hare procedure fails to satisfy monotonicity, sequential pairwise voting with a fixed agenda fails to satisfy the Pareto condition, and only a dictatorship (among those considered here) satisfied both independence of irrelevant alternatives and the always-a-winner condition.

In **Section 1.7** we gave a concrete preview of some inherent difficulties when dealing with three or more alternatives by proving that it is impossible to find a social choice procedure that satisfies the always-a-winner condition, independence of irrelevant alternatives, and the Condorcet winner criterion. Finally, we concluded in **Section 1.8** with a brief look at approval voting.

EXERCISES

The purpose of the first two exercises is to help the reader gain some familiarity with the idea of a "function from a set X to a set Y" (as defined in **Section 1.1**). For a procedure to be a function from X to Y, it must assign to each object in X and unique object in Y. In each of the following, sets X and Y are specified as is a procedure. Determine if the given procedure is or is not a function from X to Y.

1. Let X and Y both be the set of non-negative integers: $0, 1, 2, 3, \ldots$
    (a) The procedure corresponding to taking the square root of the input.
    (b) The procedure corresponding to doubling the input.
    (c) The procedure that, given input x, outputs y if and only if y is two units away from x on the number line.
    (d) The procedure that, given input x, outputs the number 17.
2. Let X be the set of (finite) non-empty sequences of nonnegative integers and let y be the set of nonnegative integers.
    (a) The procedure that, given a finite sequence, outputs y if and only if y is the seventh term of the sequence.
    (b) The procedure that outputs n if and only if n is twice the length of the sequence.
    (c) The procedure that outputs y if and only if y is greater than the last term of the sequence.
    (d) The procedure that outputs the number 17 regardless of the input.
3. For each of the six social choice procedures described in this chapter, calculate the social choice or social choices resulting from the following sequence of individual preference lists. (For sequential pairwise voting, take the agenda to be *abcde*. For the last procedure,

take the fourth person to be the dictator.)

c	d	c	b	e	d	c
a	a	e	d	d	e	a
e	e	d	a	a	a	e
b	c	a	e	c	b	b
d	b	b	c	b	c	d

4. For each of the six social choice procedures described in this chapter, calculate the social choice or social choices resulting from the following sequence of individual preference lists. (For sequential pairwise voting, take the agenda to be *acdeb*. For the last procedure, take the fifth person to be the dictator.)

a	b	c	d	e
b	c	b	c	d
e	a	e	a	c
d	d	d	e	a
c	e	a	b	b

5. Given the following sequence of individual preference lists, determine the social choice or social choices under each of the listed procedures. The horizontal line in each voter's preference list represents the cut-off line for approval voting; that is, the voter approves of each candidate above that line.

a	a	b	b	c	a	d
d	c	d	a	b	b	b
b	d	c	d	d	d	c
c	b	a	c	a	c	a

(a) Condorcet's method
(b) plurality
(c) Borda count
(d) Hare system
(e) approval voting
(f) The procedure defined as follows: If there is a Condorcet winner, that candidate is the social choice. Otherwise, use plurality to determine the social choice.

6. Given the following sequence of individual preference lists, determine the social choice or social choices under each of the listed procedures. The horizontal line in each voter's preference list represents the cut-off line for approval voting; that is, the voter approves of each candidate above that line.

Voters 1–2	Voters 3–5	Voter 6	Voters 7–8	Voters 9–13	Voters 14–17	Voter 18
a	a	b	b	c̲	c̲	c
d̲	c	d	a̲	b	b	b
b	d̲	c̲	d	d	a	a
c	b	a	c	a	d	d

    (a) Condorcet's method
    (b) plurality
    (c) Borda count
    (d) Hare system
    (e) approval voting
    (f) The procedure defined as follows: If there is a Condorcet winner, that candidate is the social choice. Otherwise, use the Hare system to determine the social choice.

7. Given the following sequence of individual preference lists, determine the social choice or social choices under each of the listed procedures. The horizontal line in each voter's preference list represents the cut-off line for approval voting; that is, the voter approves of each candidate above that line.

a	a̲	b	b	b̲	b	d	d	d
b̲	d	c	a	d	a	a	a̲	c
c	c	d̲	c̲	c	d	c	b	b̲
d	b	a	d	a	c	b	c	a

    (a) Condorcet's method
    (b) plurality
    (c) Borda count
    (d) Hare system
    (e) approval voting

(f)   The procedure defined as follows: If there is a Condorcet win-
      ner, that candidate is the social choice. Otherwise, use the
      Borda count to determine the social choice.

8.   Prove or disprove each of the following:
     (a)   Plurality voting always yields a unique social choice.
     (b)   The Borda count always yields a unique social choice.
     (c)   The Hare system always yields a unique social choice.
     (d)   Sequential pairwise voting with a fixed agenda always yields a
           unique social choice.
     (e)   A dictatorship always yields a unique social choice.

9.   Consider the following sequence of preference lists:

b	c	c	a
a	a	d	d
d	b	b	b
c	d	a	c

     (a)   Find the social choice using the Borda count.
     (b)   Suppose we change the way we assign points so that first place
           is worth 8 points, second place is worth 4 points, third place
           is worth -4 points, and fourth place is worth -8 points. Redo
           the Borda procedure using these new numbers.
     (c)   Redo (b) using the points $-1, -5, -9, -13$ for (respectively)
           first, second, third, and fourth place.
     (d)   Do as in (c) using 9, 4, 1, and 0 points for (respectively) first,
           second, third, and fourth place.
     (e)   Propose a condition on the way points are assigned that is
           sufficient to guarantee that the winner is the same as the Borda
           winner with points assigned in the usual way.

10. If we have a sequence of individual preference lists, and r and s
     are two of the alternatives, then "Net($r > s$)" is defined to be
     the number of voters who prefer r to s minus the number of
     voters who prefer s to r. Let's also change the way we assign
     points in computing the Borda score of an alternative so that
     these scores are symmetric about zero. That is, for three alter-
     natives, first place will be worth 2 points, second place will
     be worth 0 points, and third place will be worth -2 points.
     We will let "$B(r)$" denote the Borda score of the alternative r

computed using these points. Consider the following preference lists:

$$
\begin{array}{ccc}
x & x & y \\
y & z & z \\
z & y & x
\end{array}
$$

(a) Evaluate $Net(x > y)$, $Net(x > z)$, $B(x)$, and $B(z)$.
(b) Prove each of the following for this example:

$$Net(x > y) + Net(x > z) = B(x).$$
$$Net(y > z) + Net(y > x) = B(y).$$
$$Net(z > x) + Net(z > y) = B(z).$$

11. Suppose we have a social choice procedure that satisfies monotonicity. Suppose that for the four alternatives a, b, c, d we have a sequence of individual preference lists that yields d as the social choice. Suppose person one changes his list:

$$
\text{from: } \begin{array}{c} a \\ b \\ c \\ d \end{array} \quad \text{to: } \begin{array}{c} d \\ a \\ b \\ c \end{array}.
$$

Show that d is still the social choice, or at least tied for such. (The procedure one uses to show this is called "iterating the definition.")

12. Suppose we have three voters and four alternatives and suppose the individual preference lists are as follows:

$$
\begin{array}{ccc}
a & c & b \\
b & a & d \\
d & b & c \\
c & d & a
\end{array}
$$

Show that if the social choice procedure being used is sequential pairwise voting with a fixed agenda, and, if you have agenda setting power (i.e., you get to choose the order), then you can arrange for whichever alternative you want to be the social choice.

13. Consider the following social choice procedure: if there is a Condorcet winner, then that alternative is the social choice; otherwise use the Borda count to determine the social choice.
    (a) Explain why the social procedure above satisfies the Condorcet winner criterion.
    (b) Does this social choice procedure satisfy Pareto?
    (c) Does this social choice procedure satisfy monononicity?
    (d) Does this social choice procedure satisfy independence of irrelevant alternatives?

14. Prove that for a given social choice procedure and a given sequence of individual preference lists, a Condorcet winner, if it exists, must be unique.

15. Show that, for a fixed sequence of individual preference lists and an odd number of voters, an alternative is a Condorcet winner if and only if it emerges as the social choice in sequential pairwise voting with a fixed agenda regardless of the agenda.

16. Do you think there was a Condorcet winner in the 2000 Presidential Election? Explain.

17. In the 1988 Minnesota gubernatorial election, the Republican candidate Norm Coleman received 34% of the vote, Democrat Hubert "Skip" Humphrey III received 28% of the vote, and Reform candidate and professional wrestler Jesse Ventura received 37% of the vote. The election was conducted with plurality, so Jesse Ventura was the winner, yet there is evidence to suggest that most voters who voted for Coleman or Humphrey ranked Ventura third. If this is true, then 62% of the voters would have preferred either of the the other two candidates to Ventura; in other words, Ventura would have lost in a one-on-one contest against either of his opponents.

    Say that an alternative is a *Condorcet loser* if it would be defeated by every other alternative in the kind of one-on-one contest that takes place in sequential pairwise voting with a fixed agenda. Further, say that a social choice procedure satisfies the *Condorcet loser criterion* provided that a Condorcet loser is never among the social choices. Does the Condorcet loser criterion hold for:
    (a) pluality voting?
    (b) the Borda count?
    (c) the Hare system?
    (d) sequential pairwise voting with a fixed agenda?

(e) a dictatorship?

18. Prove or disprove: A Condorcet loser, if it exists, is unique.

19. Prove that for three alternatives and an arbitrary sequence of individual preference lists, there is no Condorcet loser if and only if for each alternative there is an agenda under which that alternative wins in sequential pairwise voting. (Your proof should not involve producing three particular preference lists.)

20. Modify the individual preference lists from the voting paradox to show that an alternative that loses in sequential pairwise voting for every agenda need not be a Condorcet loser. (Notice that this does not contradict Exercise 19).

21. Consider the following social choice procedure. If there is a Condorcet winner, it is the social choice. Otherwise, the alternative on top of the first person's list is the social choice. (That is, if there is no Condorcet winner, then person one acts as a dictator.) Give an example with three people and three alternatives showing that this procedure does not satisfy independence of irrelevant alternatives. (Hint: Start with the same sequence of lists that produces the voting paradox, and then move one alternative that should be irrelevant to the social choice.)

22. Prove that the social choice procedure described in Exercise 21 satisfies the Pareto condition.

23. An interesting variant of the Hare procedure was proposed by the psychologist Clyde Coombs. It operates exactly as the Hare system does, but instead of deleting alternatives with the fewest first place votes, it deletes those with the most last place votes. (In all other ways, it operates as does the Hare procedure.)
    (a) Find the social choice according to the Coombs procedure that arises from the individual preference lists in Exercise 3.
    (b) Does the Coombs system satisfy the Pareto condition?
    (c) Does the Coombs system satisfy the Condorcet winner criterion?
    (d) Does the Coombs system satisfy monotonicity?
    (e) Does the Coombs system satisfy independence of irrelevant alternatives?

24. Suppose we have two voters and three alternatives. Find preference lists so that one of the alternatives emerges as the social choice

under the Coombs procedure, but the other two emerge as tied for the win under the Hare procedure.

25. The following social choice procedure is due to A.H. Copeland. We define the *win-loss record* for an alternative to be the number of strict wins against other alternatives in a head-to-head competition minus the number of strict losses. For example, the win-loss record of a Condorcet winner is equal to one less than the total number of alternatives. Under Copeland's procedure, an alternative is a winner if no alternative has a strictly better win-loss record.
   (a) Prove that Copeland's procedure satisfies monotonicity.
   (b) Prove that Copeland's procedure satisfies the Condorcet winner criterion.

26. Say that a social choice procedure satisfies the "top condition" provided that an alternative is never among the social choices unless it occurs on top of at least one individual preference list. Prove or disprove each of the following:
   (a) Plurality voting satisfies the top condition.
   (b) The Borda count satisfies the top condition.
   (c) The Hare system satisfies the top condition.
   (d) Sequential pairwise voting satisfies the top condition.
   (e) A dictatorship satisfies the top condition.
   (f) If a procedure satisfies the top condition, then it satisfies the Pareto condition.

27. Suppose S is some social choice procedure. (Think of S as being plurality voting for the moment.) Using S we will create a new social choice procedure, denoted S^*, as follows:

> Given a sequence of preference lists $L_1, \ldots, L_n$ as input, we first reverse each of the lists to get $L_1^*, \ldots, L_n^*$.
> We now apply S to the reversed lists $L_1^*, \ldots, L_n^*$.
> These "winners" are then deleted, and the process is repeated.

If we modify the above by declaring an alternative to be the winner if—at any time in the above process—it occurs on top of at least half of the nonreversed lists, then we shall denote the system by $S\#$.
   (a) Let P denote plurality voting and apply P^* and $P\#$ to the sequence of preference lists in Exercise 3.
   (b) In a few sentences, explain why P^* is the Coombs procedure.

(c) Let H be the procedure wherein the social choice is a tie among all the alternatives except the one (or ones) having the most last place votes. Identify H^* and explain your answer in a few sentences.

(d) Make up two good questions (not involving particular lists) about the "$*$-process" or the "#-process." Don't answer the questions (unless you want to). The point of this exercise is simply to realize that good theorems are answers to good questions, and so the asking of good questions is both important and nontrivial.

28. According to the *anti-plurality* social choice procedure, the social choice is the alternative(s) with the fewest last-place rankings. Prove whether or not the anti-plurality procedure satisfies each of the following criteria.

(a) Pareto

(b) monotonicity

(c) independence of irrelevant alternatives

29. *Cumulative voting* is a social choice procedure similar to approval voting in that each voter may vote for more than one candidate. Specifically, each voter is given a set number of points C to distribute among the candidates. The candidates with the most points win. Typically, in multi-winner elections, C is equal to the number of candidates to be elected. For example, suppose that there are three open seats on the local school board. Voters have 3 points that they can distribute among the candidates; they might choose to spread out their points among two or three different candidates, or they might concentrate all three points on a single candidate. Cumulative voting is often touted as a method for helping to elect candidates from minority groups.

(a) Write down a modification of the Pareto condition that makes sense for cumulative voting.

(b) Does cumulative voting satisfy your modified version of the Pareto condition?

30. We say that a social choice system satisfies the criterion of *unanimity* if an alternative is the unique social choice whenever every ballot has that alternative ranked highest. Prove that if a social choice procedure satisfies Pareto, then it must also satisfy unanimity.

31. We say that a social choice system satisfies the criterion of *non-imposition* if every alternative occurs as the unique winner for at least one set of ballots.
    (a) Explain why non-imposition is a desirable characteristic of a social choice procedure.
    (b) Prove that if a social choice procedure satisfies unanimity, then it must also satisfy non-imposition.
32. We saw that sequential pairwise voting does not satisfy Pareto. Does it satisfy unanimity? Non-imposition?
33. During the 1995 women's World Figure Skating Championships, with only one skater left to compete, Chen Lu of China was in first place, Nicole Bobek of the U.S. was in second place, and Surya Bonaly of France was in third place. After the last skater (a relatively unknown Michelle Kwan) skated, however, the final rankings were: Chen Lu in first, Surya Bonaly in second, and Nicole Bobek in third. This example illustrates that the scoring procedure used at the time, the best of majority (BOM) method, does not satisfy IIA.

    The BOM Method works as follows. Suppose that there are six skaters. Each of say nine judges ranks the skaters. To each skater, we assign a sequence of numbers $(x_1, x_2, x_3, x_4, x_5, x_6)$ where x_i denotes the number of i-th place rankings received by the skater. Thus in the case of six skaters, we have $0 \le x_i \le 6$. Next, we compute for each skater the *lowest majority rank* (LMR): the smallest rank for which five (in general, a majority) or more judges rank that skater at that rank or higher. The *size of lowest majority* (SLM) is the number of judges that comprise this majority. For example, suppose that the judges rankings of skaters is given by:

	Judges								
	1	2	3	4	5	6	7	8	9
	1	3	2	1	5	1	1	3	1
	3	2	3	3	1	3	3	1	3
Skaters	5	1	1	2	2	5	5	2	2
	2	5	4	5	4	2	2	6	5
	6	4	6	4	3	6	6	4	6
	4	6	5	6	6	4	4	5	4

In this example, the LMR for skater 1 is 1 since a majority of judges rank her first. The SLM for skater 1 is 5 since five judges rank skater 1 first. The LMR for skater 3 is 2 since only two judges rank skater 3 first, but a majority rank her second. The SLM for skater 3 is 8 since six judges rank her second, and two rank her first.

To arrive at the final ranking of the skaters, we rank the skaters in order of ascending LMR; in the case of a tie, the skater with the higher SLM wins that rank. If there is a tie between LMR and SLM, then for each tied skater, we compute the sum of all the judges' rankings, and the skater with the smaller sum wins the rank. If there is still a tie, then the skaters remain tied in the final rankings.

(a) Calculate the LMR and SLM for skaters 3 through 6 in the example above.

(b) According to the BOM Method, rank the skaters in the example above.

(c) We saw that the BOM Method does not satisfy IIA. Does it satisfy the Pareto condition?

(d) Does the BOM method satisfy monotonicity?

(e) Does the BOM method satisfy the Condorcet winner criterion?

34. Another desirable property of social choice procedures is the *intensity of binary preferences (IBP) criterion*, introduced by Donald Saari. For a given voter's preference list, and for two alternatives x and y with x ranked higher than y, we say that the *intensity* of x over y is one more than the number of alternatives ranked between x and y. The intensity of y over x is the negative of the intensity of x over y. We say that a social choice procedure satisfies IBP provided that the following holds for every pair of alternatives x and y:

   If the social choice set includes x but not y, and one or more voters change their preferences, but no one changes the intensity of x over y in his preference list, then the social choice set should not change so as to include y.

   (a) Prove that if a social choice procedure satisfies IIA, then it must also satisfy IBP. Thus we can think of the IBP criterion as a weaker property than IIA. IBP was introduced in an attempt to explain why so many social choice procedures fail to satisfy IIA; perhaps it is not enough information to know that a voter prefers alternative x to alternative y, we might also need to know how much a voter prefers x to y.

(b)   Prove that the Borda count satisfies IBP.

(c)   Prove that a dictatorship satisfies IBP.

(d)   Prove that the Hare system does not satisfy IBP.

(e)   Does the procedure in Exercise 13 satisfy IBP? Explain.

35.   Because approval voting is very different from the social procedures
we studied earlier, not all of the desirable properties we used to evaluate the procedures even apply. We can modify them somewhat,
however, to analyze approval voting more closely. For example, we
could define a notion similar to the Pareto condition as follows:

   If every voter approves of alternative a, then a is a social choice.

(a)   Prove that with this modification, approval voting does satisfy
the Pareto condition.

A slightly different modification of the original Pareto condition is
the following:

   *If every voter prefers alternative a to alternative b, then b is not a
   social choice.*

This version of Pareto is more similar to the original than the one
mentioned above, but with the presence of the word "prefer," we are
implicitly assuming that a voter has a ranked preference list of alternatives (which may include ties), but has the power under approval
voting to draw the cutoff line between approval and disapproval at
any point in the list.

(b)   Prove that with this modification, approval voting does not
satisfy the Pareto condition.

(c)   Write down a modified version of the monotonicity condition
that makes sense for approval voting. Does approval voting
satisfy this modified monotonicity condition?

36.   Kate and her friends are planning a vacation for spring break. She
has never been to Paris, and is hoping to convince her friends that
that is the place to go. Most of the others in the group think Paris
is a good choice and rank it as their second choice, but Paris does
not have a lot of first-place votes. A few people are pushing the
Caribbean for a beach vacation, and a few others are strongly in
favor of a camping trip in the mountains. Kate has a pretty good idea
of her friends' preferences. What social choice procedure should
she suggest that the group use in order to get the trip to Paris? Why?

37.   A group of several friends are deciding which DVD to rent. At first
they are deciding between *Raiders of the Lost Arc* and *Annie Hall*.

More people are in the mood for Woody Allen than Harrison Ford, so they decide to rent *Annie Hall*. Then Michael remembers that the library now has *Breakfast at Tiffany's*, and proposes this to the group. After some discussion, the group decides to rent *Breakfast at Tiffany's*. But Riki has been racking her brain trying to think of a movie that she saw and loved a few years ago; at this point in the discussion she remembers the name: *Eternal Sunshine of the Spotless Mind*. At this point, everyone agrees to rent *Eternal Sunshine*, and they all go to the library. Which of the social choice procedures discussed in this chapter does this scenario imitate? Explain.

38. Exhibit the preference orderings that correspond to a voting paradox wherein we have ten alternatives and ten (groups of) people.

39. Construct a real-world example (perhaps involving yourself and two friends) where the individual preference lists for three (or more) alternatives are as in the voting paradox.

40. Complete the proof of the impossibility theorem in **Section 1.7** by establishing the third claim. (Your proof will be almost word-for-word the same as the proofs of Claims 1 and 2.)

41. In the 2000 presidential election, the popular vote totals were:

    George W. Bush: 50,456,002
    Al Gore: 50,999,897
    Ralph Nader: 2,882,995
    (source: Federal Election Commission)

    In the absence of the Electoral College, discuss what the outcome may have been if the Borda count or Hare system was used. Be sure to list any assumptions that you are making about voter preferences.

42. A particular voter has transitive preferences: if he or she prefers alternative a to b and b to c, then he or she also prefers a to c. One might then expect that a society's preferences are transitive as well: if the voters collectively prefer (as demonstrated by a one-on-one contest) alternative a to b and b to c, then they prefer alternative a to c. Give an example of a profile where the society's preferences are not transitive.

43. The following exercise was suggested by William Zwicker. Let's say that P_1 and P_2 are disjoint profiles if no voter is in both P_1 and P_2. For example, P_1 might be the following profile (involving voters 1–17):

Voters 1–8 Voters 9–11 Voters 12–13 Voters 14–17

a	b	c	d
b	c	b	a
c	d	d	c
d	a	a	b

and P_2 might be the following profile (involving voters 18–34):

Voters 18–25 Voters 26–28 Voters 29–30 Voters 31–34

a	c	b	d
c	b	c	a
b	d	d	b
d	a	a	c

When P_1 and P_2 are disjoint in this way, P_1+P_2 represents the combined election. Thus, in our example above, $P_1 + P_2$ has 34 voters.

A social choice function is *weakly consistent* if for every pair of disjoint profiles P_1 and P_2 (for the same set of alternatives), if an alternative x is among the winners in the P_1 election and among the winners in the P_2 election, then it is also among the winners in the $P_1 + P_2$ election.

(a) Use the above profiles to prove that the Hare system is not weakly consistent.

(b) Using a few sentences, prove that the plurality procedure is weakly consistent.

(c) Using a few sentences, prove that the Borda count is weakly consistent.

J. Smith and H. P. Young used a stronger form of consistency to help characterize an important class of voting rules called *scoring rules*. The plurality procedure and the Borda count are each scoring rules.

44. Essay Question: Arrow's theorem proves that there is no perfect social choice procedure, but some are certainly better than others. Which social choice procedure do you believe the United States should use to elect its president (within the framework of the Electoral College)?

CHAPTER

2

Yes–No Voting

■ 2.1 INTRODUCTION

In Chapter 1, we considered voting systems in which the voters were choosing among several candidates or alternatives. In the present chapter, we deal with quite a different voting situation—the one in which a single alternative, such as a bill or an amendment, is pitted against the status quo. In theses systems each voter responds with a vote of "yea" or "nay." A *yes–no voting system* is simply a set of rules that specifies exactly which collections of "yea" votes yield passage of the issue at hand.

We begin in **Section 2.2** with four real-world examples of yes–no voting systems, including the all-important U. S. federal system. In **Section 2.3** we introduce the notion of a "weighted system." These are important because they are the ones that we feel we understand (in some sense) the best. We show that—surprisingly—the voting system used in the U. N. Security Council is a weighted system. This result suggests the tantalizing possibility that *all* yes–no voting systems are weighted.

In **Section 2.4** we show that, alas, the U. S. federal system is *not* a weighted voting system. We do this via a notion known as swap robustness. We show that every weighted system is swap robust, but the U. S. federal system is not swap robust. In **Section 2.5** we generalize the notion of swap robustness to something called trade robustness, and we show that the 1982 procedure to amend the Canadian constitution is swap robust but not trade robust. We conclude in **Section 2.6** by stating the characterization theorem asserting that trade robustness and weightedness are fully equivalent.

For a monograph-length treatment of yes–no voting systems, see the 1999 book *Simple Games* by Alan Taylor and William Zwicker.

■ 2.2 FOUR EXAMPLES OF YES–NO VOTING SYSTEMS

Example 1: The European Economic Community

In 1958, the Treaty of Rome established the existence of a yes–no voting system called the European Economic Community. The voters in this system were the following six countries:

France

Germany

Italy

Belgium

the Netherlands

Luxembourg.

France, Germany, and Italy were given four votes each, while Belgium and the Netherlands were given two votes and Luxembourg one. Passage required a total of at least twelve of the seventeen votes. The European Economic Community was altered in 1973 with the addition of new countries and a reallocation of votes. This version of the European Economic Community is discussed later.

Example 2: The United Nations Security Council

The voters in this system are the fifteen countries that make up the Security Council, five of which (China, England, France, Russia, and the United States) are called permanent members whereas the other ten are called nonpermanent members. Passage requires a total of at least nine of the fifteen possible votes, subject to a veto due to a nay vote from any one of the five permanent members. (For simplicity, we ignore the possibility of abstentions.)

Example 3: The United States Federal System

There are 537 voters in this yes–no voting system: 435 members of the House of Representatives, 100 members of the Senate, the vice president, and the president. The vice president plays the role of tiebreaker in the Senate, and the president has veto power that can be overridden by a two-thirds vote of both the House and the Senate. Thus, for a bill to pass it must be supported by either:

1. 218 or more representatives and 51 or more senators (with or without the vice president) and the president.

2. 218 or more representatives and 50 senators and the vice president and the president.

3. 290 or more representatives and 67 or more senators (with or without either the vice president or the president).

Example 4: The System to Amend the Canadian Constitution

Since 1982, an amendment to the Canadian Constitution becomes law only if it is approved by at least seven of the ten Canadian provinces subject to the proviso that the approving provinces have, among them, at least half of Canada's population. For our purposes, it will suffice to work with the following population percentages for the ten Canadian provinces, based on *Statistics Canada* estimates as of January 1, 2007 (rounded):

Prince Edward Island (0%)

Newfoundland (2%)

New Brunswick (2%)

Nova Scotia (3%)

Saskatchewan (3%)

Manitoba (4%)

Alberta (11%)

British Columbia (13%)

Quebec (23%)

Ontario (39%)

In a yes–no voting system, any collection of voters is called a *coalition*. A coalition is said to be *winning* if passage is guaranteed by yes votes from exactly the voters in that coalition. Coalitions that are not winning are called *losing*. Thus, every coalition is either winning or losing. In Example 1, the coalition made up of France, Germany, and Italy is a winning coalition, as is the coalition made up of France, Germany, Italy, and Belgium. Note that when one asserts that a collection of voters is a winning coalition, nothing is being said about *how* these players actually voted on a particular issue. One is simply saying that *if* these people voted for passage of some bill and the other players voted against passage of that bill, the bill would, in fact, pass.

For most yes–no voting systems, adding extra voters to a winning coalition again yields a winning coalition. Systems with this property are said to be *monotone*. For monotone systems, one can concentrate on the so-called *minimal winning coalitions*: those winning coalitions with the property that the deletion of one or more voters from the coalition yields a losing coalition. In our example above, France, Germany, and Italy make up a minimal winning coalition, while France, Germany, Italy, and Belgium do not. (In the European Economic Community, the minimal winning coalitions are precisely the ones with exactly twelve votes—see Exercise 2.)

■ 2.3 WEIGHTED VOTING AND THE U.N. SECURITY COUNCIL

The four examples of yes–no voting systems in the last section suggest there are at least three distinct ways in which a yes–no voting system can be described:

1. One can specify the number of votes each player has and how many votes are needed for passage. This is what was done for the European Economic Community. More generally, if we start with a set of voters, then we can construct a yes–no voting system by assigning real number *weights* to the voters (allowing for what we might think of as either a fractional number of votes for some voter, or even a negative number of votes) and then set any real number q as the "quota." A coalition is then declared to be winning precisely when the sum of the weights of the voters who vote "yea" meets or exceeds the quota. (Even in the real world, quotas can be less than half the sum of the weights—see Exercise 1.)

2. One can explicitly list the winning coalitions, or, if the system is monotone, just the minimal winning coalitions. This is essentially what is done in the description of the U.S. federal system above, since the three clauses given there describe the three kinds of winning coalitions in the U.S. federal system. In fact, if one deletes the parenthetical clauses and the phrase "or more" from those descriptions, the result is a description of the three kinds of minimal winning coalitions in the U.S. federal system.

3. One can use some combination of the above two, with provisos that often involve veto power. Both the U.N. Security Council and the procedure to amend the Canadian Constitution are described in this way. Moreover, the description of the U.S. federal system in terms of the tie-breaking vote of the vice president, the presidential veto, and the Congressional override of this veto is another example of a description mixing weights with provisos and vetoes. (We say "weights" in this context since,

for example, one can describe majority support in the House by giving each member of the House weight one and setting the quota at 218, and two-thirds support from the House by setting the quota at 290.)

The following observation is extremely important for everything we shall do in this chapter: A given yes–no voting system typically can be described in more than one way.

For example, instead of using weights, we could have described the European Economic Community by listing the fourteen winning coalitions. A similar comment applies to the U.N. Security Council. In fact, *every* yes–no voting system can be described by simply listing the winning coalitions. Conversely, any collection of subsets of voters gives us a yes–no voting system, although most of the systems arrived at in this way would be of little interest.

These observations lead to the central definition of this chapter.

DEFINITION. A yes–no voting system is said to be a *weighted system* if it can be described by specifying real number weights for the voters and a real number quota—with no provisos or mention of veto power—such that a coalition is winning precisely when the sum of the weights of the voters in the coalition meets or exceeds the quota.

Example 1 (not surprising): The European Economic Community is a weighted voting system.

There is no doubt about this, since we described the system by explicitly producing the weights and the quota. Consider, however:

Example 2 (surprising): The U.N. Security Council is also a weighted voting system.

This is not obvious. Although our description of this voting system did involve weights (weight one for each of the fifteen members) and a quota (nine), it also involved the statement that each of the five permanent members has veto power.

So how does one show that a given yes–no voting system is, in fact, weighted? The answer is that one must find (that is, *produce)* weights for each of the voters and produce a quota q so that the winning coalitions are precisely the ones with weight at least q. The difficulty with doing this for the U.N. Security Council is that we must somehow assign extra weight (but how much?) to the permanent members in such a way that this weight advantage alone builds in the veto effect.

Here is the intuition behind our method of finding a set of weights and a quota q that will prove the U.N. Security Council is a weighted voting system. Since all of the nonpermanent members clearly have the same influence, we will begin by assigning them all the same weight, and we will take this weight to be 1. The five permanent members also have the same influence and so we will (temporarily) assign them all the same unknown weight x. Now, exactly what must x and q satisfy for this to work?

Consider a coalition made up of all ten nonpermanent members together with any four permanent members. This has weight $4x + 10$, and must be losing since the one permanent member not in the coalition has veto power. Thus $4x + 10 < q$. On the other hand, the five permanent members together with any four nonpermanent members is a winning coalition, and so $q \le 5x + 4$. Putting these two inequalities together yields $4x + 10 < 5x + 4$, and so $6 < x$.

The previous two paragraphs suggest that in our pursuit of weights and quota, we try weight 1 for the nonpermanent members and weight 7 for the permanent members. Our inequalities

$$4x + 10 < q \quad \text{and} \quad q \le 5x + 4$$

now imply that $38 < q \le 39$. Thus, we certainly want to try $q = 39$.

It will now be relatively easy to demonstrate that these weights and quota do, indeed, show that the U.N. Security Council is a weighted voting system. (We should also note that to prove a system is weighted, it suffices to produce the weights and quota and show they "work." It is not necessary to explain how you found them.) The argument we seek runs as follows.

PROPOSITION. *The U.N. Security Council is a weighted system.*

PROOF. Assign weight 7 to each permanent member and weight 1 to each nonpermanent member. Let the quota be 39. We must now show that each winning coalition in the U.N. Security Council has weight at least 39, and that each losing coalition has weight at most 38.

A winning coalition in the U.N. Security Council must contain all five permanent members (a total weight of 35) and at least four nonpermanent members (an additional weight of 4). Hence, any winning coalition meets or exceeds the quota of 39. A losing coalition, on the other hand, either omits a permanent member, and thus has weight at most

$$(7 \times 4) + (1 \times 10) = 28 + 10 = 38,$$

or contains at most three nonpermanent members, and thus has weight at most

$$(7 \times 5) + (1 \times 3) = 35 + 3 = 38.$$

Hence, any losing coalition falls short of the quota of 39. This completes the proof.

It turns out that if one alters any weighted voting system by giving one or more of the voters veto power, the resulting yes–no voting system is again a weighted voting system. See Exercises 5 and 6 at the end of the chapter.

A natural response to what we have done so far is to conjecture that *every* yes–no voting system is weighted. Perhaps, for example, even for the U.S. federal system, we can do something as clever as what we just did for the U.N. Security Council to find weights and a quota that work. Alas, this turns out not to be the case, as we will see in the next section.

■ 2.4 SWAP ROBUSTNESS AND THE NONWEIGHTEDNESS OF THE FEDERAL SYSTEM

The U.S. federal system, it turns out, is not a weighted voting system. But how does one prove that a system is *not* weighted? Surely we cannot simply say that we tried our very hardest to find weights and a quota and nothing that we tried appeared to work. Asserting that a system is not weighted is saying no one will *ever* find weights and a quota that describe the system. Moreover, we cannot check all possible choices of weights and quota since there are infinitely many such choices.

Here is one answer. To prove that the U.S. federal system is not weighted, it suffices to find a property that we can prove

1. holds for every weighted voting system, and

2. does not hold for the U.S. federal system.

One such property is given in the following definition:

DEFINITION. A yes–no voting system is said to be *swap robust* if a one-for-one exchange of players (a "swap") between two winning coalitions X and Y leaves at least one of the two coalitions winning. One of the players in the swap must belong to X but not Y, and the other must belong to Y but not X.

Thus, to prove that a system is swap robust, we must start with two *arbitrary* winning coalitions X and Y, and an *arbitrary* player x that is in X but not in Y, and an *arbitrary* player y that is in Y but not in X. We then let X' and Y' be the result of exchanging x and y. (Thus, x is now in Y' but not X', while y is now in X' but not Y'.) We must then show that either X' or Y' is winning. To illustrate this, consider the following.

PROPOSITION. *Every weighted voting system is swap robust.*

PROOF. Assume we have a weighted voting system and two arbitrary winning coalitions X and Y with X containing at least one voter x not in Y and Y containing at least one voter y not in X. Suppose now that voter x from the winning coalition X is exchanged for voter y from the winning coalition Y to yield X' and Y' as above. If x and y have the same weight then both X' and Y' are winning, since X' weighs the same as X and Y' weighs the same as Y. If, on the other hand, x is heavier than y, it then follows that Y' weighs strictly more than Y, since Y' was obtained by deleting y and adding the heavier x. Thus the weight of Y' certainly exceeds the quota, and thus Y' is winning as desired. (In this latter case, X' may or may not be winning.) If y is heavier than x, then the argument is analogous to what we just gave. This completes the proof.

Our goal now is to show that a particular yes–no voting system—the U.S. federal system—is *not* swap robust. To do this we must *produce* two winning coalitions X and Y and a trade between them that renders both losing. Intuitively, X and Y should both be "almost losing" (in the sense that we hope to make both actually losing by a one-for-one trade). Thus, we will try to find appropriate X and Y among the *minimal* winning coalitions. (Results related to this are in Exercise 17.)

The key to showing that the U.S. federal system is not swap robust is the following observation: if one begins with two minimal winning coalitions in the U.S. federal system and swaps a senator for a House member, then both coalitions become losing (as desired) since one of the resulting coalitions has too few senators (although a surplus of House members) and the other has too few House members (although a surplus of senators). If we simply formalize this slightly, we have:

PROPOSITION. *The U.S. federal system is not swap robust.*

PROOF. Let X consist of the president, the 51 shortest senators, and the 218 shortest members of the House. Let Y consist of the president, the 51 tallest senators, and the 218 tallest members of the House. Now let x be the shortest senator and let y be the tallest member of the House. Notice that both X and Y are winning coalitions, and that x is in X but not in Y and y is in Y but not in X. Let X' and Y' be the result of swapping x for y. Then X' is a losing coalition because it has only 50 senators, and Y' is a losing coalition because it has only 217 members of the House. Thus, the U.S. federal system is not swap robust.

Notice that a "swap" cannot involve a voter who is a member of both of the coalitions with which we begin. We avoided this in the above proof by making x the shortest senator whereas Y involved the 51 tallest senators. This is why x was definitely in X but not in Y.

An immediate consequence of the above theorem is the following corollary:

COROLLARY. *The U.S. federal system is not a weighted voting system.*

PROOF. If the U.S. federal system were weighted, then it would be swap robust by the first proposition in this section. But this would then contradict the proposition we just proved.

2.5 TRADE ROBUSTNESS AND THE NONWEIGHTEDNESS OF THE CANADIAN SYSTEM

Section 2.4 provided us with a very nice way to show that the U. S. federal system is not weighted—we simply showed that it is not swap robust. But will this technique always work? That is, if a yes–no voting system is truly not weighted, can we always *prove* it is not weighted (assuming we are clever enough) by showing that it is not swap robust?

Here is another way to ask this same question. The first proposition in **Section 2.4** asserts that *if* a yes–no voting system is weighted, *then* it is swap robust. The converse of this "if-then" statement is the following assertion (which we are not claiming is true): *if* a yes–no voting system is swap robust, *then* it is weighted. (Equivalently, if a yes–no voting system is not weighted, then it is not swap robust.) Just because an if-then statement is true, we cannot conclude that its converse is also true. (For example, the statement: "if a number is larger than 10, then it is larger than 8," is true, but the number 9 shows that its converse is false.) Thus, the question in the previous paragraph can be recast as simply asking if the converse of the first proposition in **Section 2.4** is true. If the converse were true, this would say that weightedness and swap robustness are fully equivalent in the sense that a yes–no voting system would satisfy one of the properties if and only if it satisfied the other property.

It turns out, however, that the converse of the first proposition in **Section 2.4** is false. That is, there are yes–no voting systems that are not weighted, but nevertheless *are* swap robust. Hence, we cannot prove that such a system fails to be weighted by using swap robustness as we did for the U.S. federal system. This raises the question of exactly how one shows that such a system is not weighted. Answering this question is the primary goal of this section, but we begin with the following.

PROPOSITION. *The procedure to amend the Canadian Constitution is swap robust (although we shall show later that it is not weighted).*

PROOF. Suppose that X and Y are winning coalitions in the system for amending the Canadian Constitution, and that x is a province ("voter") in X but not in Y and that y is a province in Y but not in X. Let X′ and Y′ be the result of swapping x for y. We must show that at least one of X′ and Y′ is still a winning coalition. That is, we must show that at least one of X′ and Y′ still satisfies both of the following conditions:

1.   It contains at least seven provinces.

2.   The provinces it contains represent at least half of the Canadian population.

Note, however, that both X′ and Y′ certainly satisfy condition 1, since each of the two coalitions started with at least seven provinces, and we simply did a one-for-one swap of provinces to obtain X′ and Y′. The rest of the argument is now reminiscent of the proof that a weighted system is swap robust. That is, if x has more population than y, then Y′ is a winning coalition since it has more population than Y, and so it satisfies condition 2 since Y satisfied condition 2. If, on the other hand, y has more population than x, then X′ is a winning coalition by an analogous argument. This completes the proof.

Our parenthetical remark in the statement of the above proposition promised a proof that the procedure to amend the Canadian Constitution is not weighted. But how do we do this? The answer lies in finding a stronger property than swap robustness that, like swap robustness, holds for every weighted voting system but that does not hold for the procedure to amend the Canadian Constitution. One such property that naturally suggests itself is the following strengthening of swap robustness:

DEFINITION. A yes–no voting system is said to be *trade robust* if an arbitrary exchange of players (that is, a series of trades involving groups of players) among several winning coalitions leaves at least one of the coalitions winning.[†]

[†]There are some subtleties here that need not concern a student. For example, we really are working with *sequences* of coalitions instead of *sets* of coalitions. For a thorough treatment, see Taylor-Zwicker, 1999.

Thus, trade robustness differs from swap robustness in two important ways:

1. In trade robustness, the exchanges of players are not necessarily one-for-one as they are in swap robustness.

2. In trade robustness, the trades may involve more than two coalitions.

The following is the expected strengthening of the first proposition in **Section 2.4**.

PROPOSITION. *Every weighted voting system is trade robust.*

PROOF. Notice that a series of trades among several winning coalitions leaves the number of coalitions to which each voter belongs unchanged. Thus, the total weight of all the coalitions added together is unchanged. Moreover, since the total number of coalitions is also unchanged, it follows that the average weight of these coalitions is unchanged as well.

Thus, if we start with several winning coalitions in a weighted voting system, then all of their weights at least meet quota. Hence, their average weight at least meets quota. After the trades, the average weight of the coalitions is unchanged and so it still at least meets quota. Thus, at least one of the coalitions must itself meet quota (since the average of a collection of numbers each less than quota would itself be less than quota). Hence, at least one of the coalitions resulting from the trades is winning, as desired.

To conclude that the system to amend the Canadian Constitution is not weighted, it suffices to establish the following:

PROPOSITION. *The procedure to amend the Canadian Constitution is not trade robust.*

PROOF. Let X and Y be the following winning coalitions (with percentages of population residing in the provinces also given):

X	Y
Prince Edward Island (0%)	New Brunswick (2%)
Newfoundland (2%)	Nova Scotia (3%)
Manitoba (4%)	Manitoba (4%)
Saskatchewan (3%)	Saskatchewan (3%)
Alberta (11%)	Alberta (11%)
British Columbia (13%)	British Columbia (13%)
Quebec (23%)	Ontario (39%)

Now let X' and Y' be obtained by trading Prince Edward Island and Newfoundland for Ontario. It then turns out that X' is a losing coalition because it has too few provinces (having given up two provinces in exchange for one), while Y' is a losing coalition because the eight provinces in Y' represent less than half of Canada's population.

COROLLARY. *The procedure to amend the Canadian Constitution is not a weighted voting system.*

Lani Guinier was nominated by President Clinton to be his assistant attorney general for civil rights. Her nomination was later withdrawn, in part because some of her views were considered radical. One such view involved the desirability (in certain circumstances) of a "minority veto." Interestingly, this idea also leads to a system that is swap robust but not trade robust. See Exercise 17.

■ 2.6 STATEMENT OF THE CHARACTERIZATION THEOREM

In this section, we conclude with what might be described as the "evolution of a theorem." This evolution is worth a moment's reflection, as it serves to illustrate how theorems not only answer questions, but raise new ones as well.

With the observation that the U.N. Security Council is a weighted voting system, the following question naturally suggests itself:

Is every yes–no voting system a weighted system?

The U.S. federal system, however, provides a negative answer to this question, since it is not swap robust while every weighted voting system is swap robust.

With this observation, one might be tempted to think that the only thing preventing a yes–no voting system from being weighted is a lack of swap robustness. Thus, one might conjecture that the following is true (although it turns out not to be):

A yes–no voting system is weighted if and only if it is swap robust.

The procedure to amend the Canadian Constitution, however, shows that this conjecture is false, since it is swap robust but not weighted. But the proof that this system is not weighted suggests that the intuition behind the above conjecture might have been sound, with its failure resulting from the limited kind of trades involved in the notion of swap robustness. This leads quite naturally to the notion of trade robustness, and the conjecture that trade robustness exactly characterizes the weighted voting systems. This conjecture, in fact, turns out to be true, although its proof will not be given here. The result is proven in Taylor and Zwicker (1992,1999), although its precursor goes back several decades to Elgot (1960).

THEOREM. *A yes–no voting system is weighted if and only if it is trade robust.*

It would be nice if trade robustness could be defined in terms of trades between only two coalitions. This turns out not to be the case. In Chapter 8, we consider this particular issue and others related to yes–no voting systems, and we show that if one generalizes the notions of "weights" and "quota" from numbers to vectors (defined later), then every yes–no voting system is a "vector-weighted" system.

....

■ 2.7 CONCLUSION

In this chapter we considered voting systems in which a single alternative, such as a bill or an amendment, is pitted against the status quo. Four examples of such yes–no voting systems were presented:

the European Economic Community, the U.N. Security Council, the U.S. federal system, and the 1982 procedure to amend the Canadian Constitution.

By definition, a yes–no voting system is weighted if it can be described by specifying a weight for each player and a quota so that the winning coalitions are precisely the ones whose total weight meets or exceeds the quota. (Such a description cannot involve provisos involving things like veto power.) We showed that the European Economic Community is obviously a weighted voting system, and that the U.N. Security Council is also a weighted voting system, although this is a little less obvious.

The remainder of this chapter dealt with the question of how one shows that a given yes–no voting system is not weighted. In fact, we showed that neither the U.S. federal system nor the procedure to amend the Canadian Constitution is a weighted voting system. For the former, we proved that every weighted voting system is swap robust (that is, a one-for-one swap of players between two winning coalitions leaves at least one of the resulting coalitions winning), but the U.S. federal system is not swap robust (we traded a senator for a member of the House, thus rendering two minimal winning coalitions losing). We showed that this approach, however, would not work for the procedure to amend the Canadian Constitution, since this system is, in fact, swap robust. Thus, we introduced a stronger property, called trade robustness, that we could show holds for every weighted system, but does not hold for the procedure to amend the Canadian Constitution.

From a practical point of view, this chapter has provided some fairly easy to use techniques to show that certain yes–no voting systems are not weighted. We concluded with a statement of the theorem that characterizes weighted voting systems as precisely those that are trade robust.

EXERCISES

1. One way to have a case heard by the Supreme Court of the United States involves what is called a grant of certiorari. The effect of this is that a case will be considered by the court if at least four of the nine justices deem the issue worthy of further consideration. Explain how this corresponds to a weighted yes–no voting system

with quota less than half the sum of the weights. Can this give rise to paradoxical situations? Why or why not? Can you think of similar situations where a quota of less than half the sum of the weights might make sense?

2. Explain why the minimal winning coalitions in the European Economic Community are precisely the ones having exactly twelve votes. (First explain why it is that *if* a coalition has exactly twelve votes, *then* it is a minimal winning coalition. Then explain why it is that if a coalition has either fewer than twelve votes or more than twelve votes, then it is not a minimal winning coalition.)

3. Suppose you are lobbying for a resolution that is being placed before the European Economic Community. Disregarding the influence various nations have on one another through discussion and debate, how important to you is Luxembourg's support when it comes time to actually vote on the issue. ("Not very important" is not a very good answer.)

4. Suppose the U.N. Security Council had eight nonpermanent members and three permanent members with passage requiring a total of seven votes subject to the veto power of each of the permanent members. Prove that this is a weighted voting system. (Include a discussion—as in the text—showing how you found the appropriate weights and quota.)

5. Suppose we have a four-person weighted voting system with positive weights a, b, c, and d for the voters named A, B, C, and D, respectively. Assume the quota is the number q. Now suppose we create a new yes–no voting system by adding a clause that gives voter A veto power. Show that this is also a weighted voting system. (Hint: Let $r = a+b+c+d$ and then take the weights to be $a+r, b, c,$ and d. Set the quota at $q + r$.)

6. Show that if a weighted voting system is altered by giving veto power to some of the voters, then the resulting yes–no voting system is again a weighted voting system. (To get started notationally, assume the voters are named $v_1, \ldots, v_k, p_{k+1}, \ldots, p_n$, with weights $w_1, \ldots, w_n$ and suppose we want to give veto power to $v_1, \ldots, v_k$.)

7. Suppose that five people—named A, B, C, D, and E—are seated at a circular table with A next to B next to C next to D next to E next to A. Suppose we form a yes–no voting system by declaring a coalition to be winning if and only if it contains at least three people seated

next to one another (for example: D, E, and A make up a winning coalition). List the minimal winning coalitions and then determine if this system is swap robust. Is it weighted?

8. The Bennet family consists of both parents, Tom and Jane, and the three children Rachel, Melissa, and Rob. When making major decisions like whether to go to Cape Cod again this year for vacation or try something new, they use the following voting system: To pass, any decision must have the support of at least one parent and two children.

   (a) List all winning coalitions.

   (b) Is this voting system swap robust? Either explain why it is in two or three sentences, or produce two winning coalitions and a swap that illustrate why it is not.

9. The math department at the local college uses the following yes–no voting system to make decisions. All assistant professors have one vote, all associate professors have three votes, and all full professors have five votes. For a proposal to pass, it must have the support of:

   (i)   at least one-third of the total number of assistant professors,

   (ii) at least half of the total number of full professors, and

   (iii) at least half of the total number of possible votes.

   (a) Suppose the department has 10 full professors, 4 associate professors, and 6 assistant professors. If the proposal to require all assistant professors to give weekend lectures at the local elementary school has the support of 7 full professors, 2 associate professors, and no assistant professors, does it pass?

   (b) Suppose the department has 4 full professors, 2 associate professors, and 2 assistant professors. List all winning coalitions.

   (c) Is the system described above swap robust?

   (d) Is the system described above a weighted voting system?

10. A board of directors for a small corporation consists of three members: Derek with 5 votes as the largest shareholder, Christina with 3 votes as the next largest shareholder, and Isabelle with 2 votes as the smallest shareholder. The quota for passage of any issue is 7 votes. Prove that although each of the three directors has a different number of votes, each has equal power in the sense that

no one voter can singlehandedly pass a motion and any two voters can pass any motion.

11. Suppose we have a "minority veto" yes–no voting system with five voters: majority voters A, B, C and minority voters R and S. Passage requires a total of at least three votes, at least one of which must be a minority voter. Is this a weighted voting system? If so, find weights for each voter and a quota.

12. Consider the version of the UN Security Council in which there are 2 non-permanent members, 4 permanent members, and passage requires at least 5 of the 6 subject to a veto by any of the 4 permanent members.
    (a) Find weights and a quota showing that this is weighted.
    (b) Show that the weights and quota you found in part (a) work.

13. Consider the version of the UN Security Council in which there are 8 non-permanent members, 3 permanent members, and passage requires at least 6 of the 11 subject to a veto by any of the 3 permanent members.
    (a) Find weights and a quota showing that this is weighted.
    (b) Show that the weights and quota you found in part (a) work.

14. Suppose the voters are the numbers 1, 2, 3, 4, and 5, and that the winning coalitions are the ones that contain at least three consecutive numbers. Determine if this is a weighted voting system.

15. Suppose we have four voters: $A_1, A_2, A_3,$ and A_4. Let S_2 denote the yes–no voting system wherein a coalition X is winning if and only if:
    (i) at least one of A_1 and A_2 is in X, and
    (ii) at least one of A_3 and A_4 is in X.
    Prove that S_2 is not weighted.

16. One of the early proposals for a system to amend the Canadian Constitution was the 1971 Victoria Charter. In this system, Ontario and Quebec were given veto power, winning coalitions had to contain at least two of the western provinces (Manitoba, Saskatchewan, Alberta, and British Columbia) representing at least half of the population of the western provinces, and they had to contain at least two of the Atlantic provinces (Newfoundland, Prince Edward Island, Nova Scotia, and New Brunswick). Prove that this is not a weighted voting system.

17. The following is a simple example (suggested by Matthew Frisoni while he was a student at Siena College) of what Lani Guinier had in

mind as a "minority veto system." Assume we have a majority group consisting of five voters: A, B, C, D, E, and a minority group consisting of three voters: F, G, and H. Assume that passage requires not only approval of at least five of the eight voters, but also approval of at least two of the three minority voters.

(a) Prove that this system is swap robust.

(b) Prove that this system is not trade robust.

18. In a yes–no voting system, a *dummy voter* is a voter whose addition to a losing coalition never renders it winning.

Sarah and Tim propose a voting system to their younger siblings by which they can all make decisions. Sarah and Tim, the 16-year old twins, will each receive 6 votes. Cindy who is half their age, gets two votes, and Andy, the 5-year old, gets 1 vote.

(a) Find a quota for the system for which there are no dummy voters.

(b) Find a quota for the system for which both Cindy and Andy are dummy voters.

(c) Find a quota for the system for which Andy alone is a dummy voter.

19. A *dictator* in a yes–no voting system is a voter who is a member of every winning coalition and is not a member of any losing coalitions.

(a) Prove that Ed is a dictator in the following yes–no voting system: Ed has 10 votes, Josh has 5 votes, Will has 4 votes, and Sean has 2 votes. The quota is 12.

(b) Is it possible for a yes–no voting system to have a dictator without any dummy voters?

20. Let's agree to call a yes–no voting system M-swap robust ("M" for "minimal") if a one-for-one exchange of players between two *minimal* winning coalitions leaves at least one of the two coalitions winning.

(a) Prove that there is an M-swap robust yes–no voting system that is not swap robust. (Hint: Let the voters be x, y, a, b, and let the winning coalitions be xab, yb, x, a, b, y.)

(b) Prove that every monotone M-swap robust yes–no voting system is swap robust. (Hint: Work with the contrapositive, that is, assume we have an arbitrary yes–no voting system that is not swap robust, and then show that it is not M-swap robust. This takes a little thought.)

21. Suppose we want to specify a notion of "winning coalition" by explic-
     itly listing those subsets that we wish to have so designated. If our
     notion is to be a reasonable one, which of the following properties
     should be satisfied?

     (a) If X is a winning coalition and every voter in X is also in Y, then
          Y is also a winning coalition.
     (b) If X and Y are winning coalitions, then so is the coalition
          consisting of voters in both X and in Y.
     (c) If X is a winning coalition and every voter in Y is also in X, then
          Y is also a winning coalition.
     (d) If X and Y are disjoint (that is, have no voters in common), then
          at least one fails to be a winning coalition.
     (e) If X and Y are winning coalitions, then so is the coalition
          consisting of voters in either X or in Y.
     (f) If X is the set of all players, then X is a winning coalition.
     (g) If X is the empty set, then X is a winning coalition.
     (h) If X is a winning coalition and X is split into two sets Y and Z
          so that every voter in X is in exactly one of Y and Z, then either
          Y is a winning coalition or Z is a winning coalition.

22. Suppose that we have a yes–no voting system with three voters
     $A, B,$ and $C,$ where the winning coalitions are as follows:

$$ABC, AB, AC, A.$$

     (a) Is there a dictator in this voting system?
     (b) Are there any dummy voters?

23. Is the yes–no voting system in the previous exercise trade robust?

24. Suppose we have a yes–no voting system with four voters $A, B, C,$
     and $D,$ where the winning coalitions are as follows:

$$ABCD, ABC, ABD, ACD, BCD, AB, AD.$$

     Is this a weighted voting system? If so, find weights for each voter
     and a quota.

25. Prove that a yes–no voting system is weighted if and only if it is pos-
     sible to assign weights to the voters so that every winning coalition
     is strictly heavier than every losing coalition.

26. Consider the 6-person voting system in which voters $A, B,$ and C belong to chamber 1, and voters $D, E,$ and F belong to chamber 2. Suppose that a coalition is winning precisely when it contains at least 2 voters from each chamber. Prove that this system is not swap robust.

27. Suppose we have a 7-person yes–no voting system with a 4-person House $H = \{a, b, c, d\}$ and a 3-person Senate $S = \{x, y, z\}$. Suppose that a coalition is winning when it has a total of at least 3 voters, at least one of which is from the Senate.

(a) Prove or disprove that this system is swap robust.

(b) Prove or disprove that this system is trade robust.

28. Suppose that a yes–no voting system is swap robust, and that we create a new yes–no voting system by giving voter p veto power (thus, a coalition is winning in the new system precisely when it is both winning in the old system and contains the voter p).

(a) Prove that the new system is swap robust.

(b) If we alter a yes–no voting system by giving 3 voters veto power, is it still swap robust (why or why not)?

29. Use trade-robustness to prove that if we have a weighted yes–no voting system, and we create a new system by giving some of the voters veto power, then the resulting system is still weighted.

CHAPTER
3

Political Power

■ 3.1 INTRODUCTION

One of the central concepts of political science is power. While power itself is certainly many-faceted (with aspects such as influence and intimidation coming to mind), our concern is with the narrower domain involving power as it is reflected in formal voting situations (most often) related to specific yes–no issues. If everyone has one vote and majority rule is being used, then clearly everyone has the same amount of "power." Intuition might suggest that if one voter has three times as many votes as another (and majority rule is still being used in the sense of "majority of votes" being needed for passage), then the former has three times as much power as the latter. The following hypothetical example should suffice to call this intuition (or this use of the word *power*) into question.

Suppose the United States approaches its neighbors Mexico and Canada with the idea of forming a three-member group analogous to the European Economic Community as set up by the Treaty of Rome in 1958. Recall that France, Germany, and Italy were given four votes each, Belgium and the Netherlands two each, and Luxembourg one vote, for a total of seventeen votes, with twelve of the seventeen votes needed for passage. Now suppose that in our hypothetical example we suggest mimicking this with the United States getting three votes while each of its two smaller neighbors gets one vote. With this total of five votes we could also suggest using majority rule (three or more out of five votes) for passage and argue that it is not unreasonable for the United States to have three times as much "power" as either Canada or Mexico. In this situation it is certainly unlikely that either Canada or Mexico would be willing to go along with the previously suggested intuition aligning "three times as many votes" with "three times as much power."

In the hypothetical example above, Canada and Mexico have no "power" (although they have votes). So what is this aspect of power that they are completely without? As an answer, "control over outcomes" suggests itself, and, indeed, much of the present chapter is devoted to quantitative measures of power that directly incorporate this control-over-outcomes aspect of power. (It also turns out—and we'll discuss this in more detail later—that these quantitative measures of power indicate that Luxembourg fared no better in the original European Economic Community of 1958 than would Canada and Mexico in our hypothetical example.)

In **Section 3.2** we consider the most well-known cardinal notion of power: the Shapley–Shubik index. This notion of power applies to any yes–no voting system (and not just to weighted voting systems). The mathematical preliminaries involved here include the "multiplication principle" and its corollary giving the number of distinct arrangements of n objects. In **Section 3.3**, we calculate the Shapley–Shubik indicies for the European Economic Community, and we use later developments in this voting system to illustrate a phenomenon known as the "paradox of new members" (where one's power is actually increased in a situation where it appears to have been diluted).

Section 3.4 contains the second most well-known cardinal notion of power: the Banzhaf index. In **Section 3.5**, we introduce two methods, dating back to Allingham (1975) and Dahl (1957), for calculating the Banzhaf index, and we illustrate these methods using both the European Economic Community and a new paradox of Felsenthal and Machover (1994) wherein a voter's power, as measured by the Banzhaf index, increases after giving away a vote. (A paradox, due to William

Zwicker, that applies to the Shapley–Shubik index but not the Banzhaf index is given in the exercises.)

Finally, in **Section 3.6** we present an aspect of political power that is quite different from those considered earlier in the chapter. The issue here is a well-known paradox called the Chair's paradox, and we use it to illustrate that naïve measures of power need not correspond to influence over outcomes.

Before continuing, we add one convention that will be in place throughout this chapter (and Chapter 9):

CONVENTION. Whenever we say "voting system" we mean "monotone voting system in which the grand coalition (the one to which all the voters belong) is winning, and the empty coalition (the one to which none of the voters belong) is losing."

With this convention at hand, we can now turn to our discussion of power.

■ 3.2 THE SHAPLEY-SHUBIK INDEX OF POWER

We begin with some mathematical preliminaries. Suppose we have n people $p_1, p_2, \ldots, p_n$ where n is some positive integer. In how many different ways (i.e., orders) can we arrange them? We check it for some small values of n in Figure 1.

Notice how the orderings for $n = 2$ in Figure 1 arise from the single ordering p_1 for $n = 1$; that is, p_2 can be placed in either the "box" before the p_1 or the "box" after the p_1, as illustrated in Figure 2.

$n = 1$: clearly only one way $\qquad p_1$

$n = 2$: two ways $\qquad p_2p_1 \quad$ and $\; / p_1p_2$

$n = 3$: six ways $\qquad p_3p_2p_1; \, p_2p_3p_1; \, p_2p_1p_3$

and

$p_3p_1p_2; \, p_1p_3p_2; \, p_1p_2p_3$

FIGURE 1

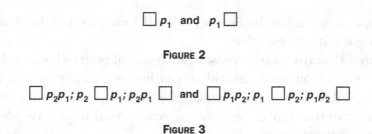

FIGURE 2

FIGURE 3

Closer analysis reveals that the same thing is happening as we go from the $n = 2$ case to the $n = 3$ case. That is, each of the two orderings of p_1 and p_2 gives rise to three orderings of p_1, p_2, and p_3 depending upon in which of the three boxes we choose to place p_3. This is illustrated in Figure 3.

If we were to display the $n = 4$ case as in Figure 1, then it should be clear that for each of the six orderings of p_1, p_2 and p_3 we'd have four boxes in which to place p_4. With four orderings so arising from each of the previous six, we would see a total of twenty-four. Examining the sequence of numbers that are suggesting themselves reveals the following:

If $n = 1$, the number of orderings is 1.
If $n = 2$, the number of orderings is $2 = 2 \times 1$.
If $n = 3$, the number of orderings is $6 = 3 \times 2 = 3 \times 2 \times 1$.
If $n = 4$, the number of orderings is $24 = 4 \times 6 = 4 \times 3 \times 2 = 4 \times 3 \times 2 \times 1$.

In general, the number of different ways that n people can be arranged (i.e., ordered) is $(n) \times (n - 1) \times (n - 2) \times \ldots \times (3) \times (2) \times (1)$. This number is called "n factorial" and is denoted by "$n!$" (e.g., $5! = 5 \times 4 \times 3 \times 2 \times 1 = 120$). All of this is formalized in the following three results.

PROPOSITION 1 *(The Multiplication Principle). Suppose we are considering objects each of which can be built in two steps. Suppose there are exactly f (for "first") ways to do the first step and exactly s (for "second") ways to do the second step. Then the number of such objects (that can be built altogether) is the product f × s. (We are assuming that different construction scenarios produce different objects.)*

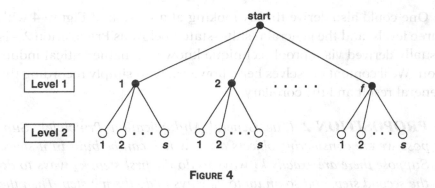

FIGURE 4

PROOF. Consider Figure 4, where the dots (from now on called *nodes*) labeled $1, 2, \ldots, f$ on the first level represent the f ways to do the first step in the construction process, and, for each of these, the nodes labeled $1, 2, \ldots, s$ represent the s ways to do the second step.

Notice that each node on level 2 (the so-called *terminal nodes*) corresponds to a two-step construction scenario. Moreover, the number of terminal nodes is clearly $f \times s$ since we have f "clumps" (one for each node on level 1) and each "clump" is of size s. This completes the proof.

Suppose now that we are building objects by a three-step process where there are k_1 ways to do the first step, k_2 ways to do the second, and k_3 ways to do the third step. How many such objects can be constructed? The answer, it turns out, can be derived from Proposition 1 because we can regard this three-step process as taking place in two "new steps" as follows:

1. New step one: same as old step one.

2. New step two: do the old step two and then the old step three.

Notice that Proposition 1 tells us there are $k_2 \times k_3$ new step twos. Since we know there are k_1 new step ones, we can apply Proposition 1 again to conclude that the number of objects built by our new two-step process (equivalently, by our old three-step process) is given by:

$$k_1 \times (k_2 \times k_3) = k_1 \times k_2 \times k_3.$$

One could also derive this by looking at a version of Figure 4 with three levels, and the general result—stated below as Proposition 2—is usually derived via a proof technique known as mathematical induction. We'll content ourselves here, however, with simply recording the general result and the corollary.

PROPOSITION 2 *(The General Multiplication Principle). Suppose we are considering objects all of which can be built in n steps. Suppose there are exactly k_1 ways to do the first step, k_2 ways to do the second step, and so on up to k_n ways to do the nth step. Then the number of such objects (that can be built altogether) is*

$$k_1 \times k_2 \times \ldots \times k_n,$$

assuming that different construction scenarios produce different objects.

As an application of Proposition 2, suppose we have n people and the objects we are building are arrangements (i.e., orders) of the people. Each ordering can be described as taking place in n steps as follows:

   Step 1: Choose one person (from the n) to be first.

   Step 2: Choose one person (from the remaining $n - 1$) to be second.

   $\vdots$

   Step $n - 1$: Choose one person (from the remaining 2) to be $n - 1^{st}$.

   Step n: Choose the only remaining person to be last.

Clearly there are n ways to do step 1, $n - 1$ ways to do step 2, $n - 2$ ways to do step 3, and so on down to 2 ways to do step $n - 1$ and 1 way to do step n. Thus, an immediate corollary of Proposition 2 is the following:

COROLLARY. *The number of different ways that n people can be arranged is*

$$n \times (n - 1) \times (n - 2) \times \cdots \times (3) \times (2) \times (1),$$

which is, of course, just $n!$ (factorial, not surprise).

One final idea—that of a "pivotal player"—is needed before we can present the formal definition of the Shapley–Shubik index. Suppose, for example, that we have a yes–no voting system with seven players: $p_1, p_2, p_3, p_4, p_5, p_6, p_7$. Fix one of the 7! orderings; for example, let's consider:

$$p_3 \; p_5 \; p_1 \; p_6 \; p_7 \; p_4 \; p_2.$$

We want to identify one of the players as being "pivotal" for this ordering. To explain this idea, we picture a larger and larger coalition being formed as we move from left to right. That is, we first have p_3 alone, then p_5 joins to give us the two-member coalition p_3, p_5. Then p_1 joins, yielding the three-member coalition p_3, p_5, p_1. And so on. The pivotal person for this ordering is the one whose joining converts this growing coalition from a non-winning one to a winning one. Since the empty coalition is losing and the grand coalition is winning (by our convention in **Section 3.1**), it is easy to see that some voter must be pivotal.

Example:

Suppose $X = \{p_1, \ldots, p_7\}$ and each player has one vote except p_4 who has three. Suppose five votes are needed for passage. Consider the ordering: $p_7 p_3 p_5 p_4 p_2 p_1 p_6$. Then, since $\{p_7, p_3, p_5\}$ is not a winning coalition, but $\{p_7, p_3, p_5, p_4\}$ is a winning coalition, we have that the pivotal player for this ordering is p_4.

The Shapley–Shubik index of a player p is the number between zero and one that represents the fraction of orderings for which p is the pivotal player. Thus, being pivotal for lots of different orderings corresponds to having a lot of power according to this particular way of measuring power. More formally, the definition runs as follows.

DEFINITION. Suppose p is a voter in a yes–no voting system and let X be the set of all voters. Then the *Shapley–Shubik index* of p, denoted here by SSI(p), is the number given by:

$$\text{SSI}(p) = \frac{\text{the number of orderings of } X \text{ for which } p \text{ is pivotal}}{\text{the total number of possible orderings of the set } X}.$$

Note the following:

1. The denominator in SSI(p) is just $n!$ if there are n voters.
2. For every voter p we have $0 \leq \text{SSI}(p) \leq 1$.
3. If the voters are $p_1, \ldots, p_n$, then $\text{SSI}(p_1) + \ldots + \text{SSI}(p_n) = 1$.

Intuitively, think of SSI (p) as the "fraction of power" that p has. The following easy example is taken from Brams (1975); it is somewhat striking.

Example:

Suppose we have a three-person weighted voting system in which p_1 has fifty votes, p_2 has forty-nine votes, and p_3 has one vote. Assume fifty-one votes are needed for passage. The six possible orderings ($3! = 3 \times 2 \times 1 = 6$) are listed below, and the pivotal player for each has been circled.

$$p_1 \quad \circled{p_2} \quad p_3$$
$$p_1 \quad \circled{p_3} \quad p_2$$
$$p_2 \quad \circled{p_1} \quad p_3$$
$$p_2 \quad p_3 \quad \circled{p_1}$$
$$p_3 \quad \circled{p_1} \quad p_2$$
$$p_3 \quad p_2 \quad \circled{p_1}$$

Since p_1 is pivotal in four of the orderings, $\text{SSI}(p_1) = \frac{4}{6} = \frac{2}{3}$.
Since p_2 is pivotal in one of the orderings, $\text{SSI}(p_2) = \frac{1}{6}$.
Since p_3 is pivotal in one of the orderings, $\text{SSI}(p_3) = \frac{1}{6}$.
Notice that although p_2 has forty-nine times as many votes as p_3, they each have the same fraction of power (at least according to this particular way of measuring power).

3.3 CALCULATIONS FOR THE EUROPEAN ECONOMIC COMMUNITY

We now return to the European Economic Community as set up in 1958 and calculate the Shapley–Shubik index for the member countries. Recall that France, Germany, and Italy had four votes, Belgium and the Netherlands had two votes, and Luxembourg had one vote. Passage required at least twelve of the seventeen votes.

Let's begin by calculating SSI(France). We'll need to determine how many of the 6! = 720 different orderings of the six countries have France as the pivotal player. Because 720 is a fairly large number, we will want to get things organized in such a way that we can avoid looking at the 720 orderings one at a time.

Notice first that France is pivotal for an ordering precisely when the number of votes held by the countries to the left of it is either eight, nine, ten, or eleven. (If the number were seven or less, then the addition of France's four votes would yield a total of at most eleven, and thus not make it a winning coalition. If the number were twelve or more, it would be a winning coalition without the addition of France.) We'll handle these four cases separately, and then just add together the number of orderings from each case in which France is pivotal to get the desired final result.

Case 1: Exactly Eight Votes Precede France

There are three ways to total eight with the remaining numbers. We'll handle each of these as a subcase.

1.1: France is Preceded by Germany, Belgium, and the Netherlands (with Votes 4, 2, and 2)

In this subcase, the three countries preceding France can be ordered in 3! = 6 ways, and for each of these six, the two countries following France (Italy and Luxembourg in this case) can be ordered in 2! = 2 ways. Thus we have 6 × 2 = 12 distinct orderings in this subcase. (Equivalently, the number of orderings in this subcase—by Proposition 2—is 3 × 2 × 1 × 1 × 2 × 1 = 12.)

1.2: France is Preceded by Italy, Belgium, and the Netherlands (with Votes 4, 2, and 2)

This case is exactly as 1.1, since both Germany and Italy have four votes.

1.3: France is Preceded by Germany and Italy (with Votes 4 and 4)

In this subcase, the two countries preceding France can be ordered in 2! = 2 ways, and for each of these two, the three countries following France (Belgium, the Netherlands, and Luxembourg in this case) can be ordered in 3! = 6 ways. Thus we have 2 × 6 = 12 distinct orderings in this subcase, also.

Hence, in case 1 we have a total of 36 distinct orderings in which France is pivotal. For the next three cases (and their subcases), we'll leave the calculations to the reader and just record the results.

Case 2: Exactly Nine Votes Precede France

2.1: France is Preceded by Germany, Belgium, the Netherlands, and Luxembourg (with Votes 4, 2, 2, and 1)

The number of orderings here turns out to be 4! × 1! = 24 × 1 = 24.

2.2: France is Preceded by Italy, Belgium, the Netherlands, and Luxembourg (with Votes 4, 2, 2, and 1)

As in 2.1, the number of orderings here is 24.

2.3: France is Preceded by Germany, Italy, and Luxembourg (with Votes 4, 4, and 1)

The number of orderings turns out to be 3! × 2! = 6 × 2 = 12.

Hence, in case 2 we have a total of 60 distinct orderings in which France is pivotal.

Case 3: Exactly Ten Votes Precede France

3.1: France is Preceded by Germany, Italy, and Belgium (with Votes 4, 4, and 2)

The number of orderings turns out to be 3! × 2! = 6 × 2 = 12.

3.2: France is Preceded by Germany, Italy, and the Netherlands (with Votes 4, 4, and 2)

Exactly as in 3.1, the number here is 12.

Hence, in case 3 we have a total of 24 distinct orderings in which France is pivotal.

Case 4: Exactly Eleven Votes Precede France

4.1: France is Preceded by Germany, Italy, Belgium, and Luxembourg (with Votes 4, 4, 2, and 1)

The number of orderings turns out to be $4! \times 1! = 24 \times 1 = 24$.

4.2: France is Preceded by Germany, Italy, the Netherlands, and Luxembourg (with Votes 4, 4, 2, and 1)

Exactly as in 4.1, the number here is 24.

Hence, in case 4 we have a total of 48 distinct orderings in which France is pivotal.

Finally, to calculate the Shapley–Shubik index of France, we simply add up the number of orderings from the above four cases (giving us the number of orderings for which France is pivotal), and divide by the number of distinct ways of ordering six countries (which is $6! = 720$). Thus,

$$\text{SSI(France)} = \frac{36 + 60 + 24 + 48}{720} = \frac{168}{720} = \frac{14}{60} \approx 23.3\%$$

Germany and Italy also have a Shapley–Shubik index of 14/60 since, like France, they have four votes. It turns out that the Netherlands and Belgium both have a Shapley–Shubik index of 9/60, although we'll leave this as an exercise (which can be done in two different ways) at the end of the chapter. Another exercise is to show that poor Luxembourg has a Shapley–Shubik index of zero! (Hint: in order for Luxembourg to be pivotal in an ordering, exactly how many votes would have to be represented by countries preceding Luxembourg in the ordering? What property of the numbers giving the votes for the other five countries makes this total impossible?) These results are summarized in the following chart:

Country	Votes	Percentage of votes	SSI	Percentage of power
France	4	23.5	14/60	23.3
Germany	4	23.5	14/60	23.3
Italy	4	23.5	14/60	23.3
Belgium	2	11.8	9/60	15.0
Netherlands	2	11.8	9/60	15.0
Luxembourg	1	5.9	0	0

We conclude this section by using the European Economic Community to illustrate a well-known paradox that arises with cardinal notions of power such as those considered in the present chapter (and later in Chapter 9). The setting is as follows: Suppose we have a weighted voting body as set up among France, Germany, Italy, Belgium, the Netherlands, and Luxembourg in 1958. Suppose now that new members are added and given votes, but the percentage of votes needed for passage remains about the same. Intuitively, one would expect the "power" of the original players to become somewhat diluted, or, at worst, to stay the same. The rather striking fact that this need not be the case is known as the "Paradox of New Members." It is, in fact, precisely what occurred when the European Economic Community expanded in 1973.

Recall that in the original European Economic Community, France, Germany, and Italy each had four votes, Belgium and the Netherlands each had two votes, and Luxembourg had one, for a total of seventeen. Passage required twelve votes, which is 70.6 percent of the seventeen available votes. In 1973, the European Economic Community was expanded by the addition of England, Denmark, and Ireland. It was decided that England should have the same number of votes as France, Germany, and Italy, but that Denmark and Ireland should have more votes than the one held by Luxembourg and fewer than the two held by Belgium and the Netherlands. Thus, votes for the original members were scaled up by a factor of $2\frac{1}{2}$, except for Luxembourg, which only had its total doubled. In summary then, the countries and votes stood as follows:

France	10	Belgium	5	England	10
Germany	10	Netherlands	5	Denmark	3
Italy	10	Luxembourg	2	Ireland	3

The number of votes needed for passage was set at forty-one, which is 70.7 percent of the fifty-eight available votes.

The striking thing to notice is that Luxembourg's power—as measured by the Shapley–Shubik index—has increased. That is, while Luxembourg's Shapley–Shubik index had previously been zero, it is clearly greater than zero now since we can produce at least one ordering of the nine countries for which Luxembourg is pivotal. (The actual production of such an ordering is left as an exercise at the end of the chapter.) Notice also that this increase of power is occurring in spite of the fact that Luxembourg was treated worse than the other countries in the scaling-up process. For some even more striking instances of this paradox of new members phenomenon, see the exercises at the end of the chapter where, for example, it is pointed out that even if Luxembourg had been left with one vote, its power still would have increased.

■ 3.4 THE BANZHAF INDEX OF POWER

A measure of power that is similar to (but not the same as) the Shapley–Shubik index is the so-called Banzhaf index of a player. This power index was introduced by the attorney John F. Banzhaf III in connection with a lawsuit involving the county board of Nassau County, New York in the 1960s (see Banzhaf, 1965). The definition takes place via the intermediate notion of what we shall call the "total Banzhaf power" of a player. The definition follows.

DEFINITION. Suppose that p is a voter in a yes–no voting system. Then the total Banzhaf power of p, denoted here by TBP(p), is the number of coalitions C satisfying the following three conditions:

1. p is a member of C.
2. C is a winning coalition.
3. If p is deleted from C, the resulting coalition is not a winning one.

If C is a winning coalition, but the coalition resulting from p's deletion from C is not, then we say that p's *defection from C is critical.*

Notice that TBP(p) is an integer (whole number) as opposed to a fraction between zero and one. To get such a corresponding fraction, we do the following (which is called "normalizing").

DEFINITION. Suppose that p_1 is a player in a yes–no voting system and that the other players are denoted by $p_2, p_3, \ldots, p_n$. Then the *Banzhaf index* of p_1, denoted here by BI(p_1), is the number given by

$$\text{BI}(p_1) = \frac{\text{TBP}(p_1)}{\text{TBP}(p_1) + \cdots + \text{TBP}(p_n)}.$$

Notice that $0 \leq \text{BI}(p) \leq 1$ and that if we add up the Banzhaf indices of all n players, we get the number 1.

Example:

Let's again use the example where the voters are p_1, p_2, and p_3; and p_1 has fifty votes, p_2 has forty-nine votes, p_3 has one vote; and fifty-one votes are needed for passage. We will calculate **TBP** and **BI** for each of the three players. Recall that the winning coalitions are

$$C_1 = \{p_1, p_2, p_3\},$$
$$C_2 = \{p_1, p_2\},$$
$$C_3 = \{p_1, p_3\}.$$

For TBP(p_1), we see that p_1 is in each of the three winning coalitions and his defection from each is critical. On the other hand, neither p_2's nor p_1's defection from C_1 is critical, but p_2's is from C_2 and p_3's is from C_3. Thus:

$$\text{TBP}(p_1) = 3 \qquad \text{TBP}(p_2) = 1 \qquad \text{TBP}(p_3) = 1$$

and, thus,

$$\text{BI}(p_1) = \frac{3}{(3+1+1)} = \frac{3}{5}$$

$$\text{BI}(p_2) = \frac{1}{(3+1+1)} = \frac{1}{5}$$

$$\text{BI}(p_3) = \frac{1}{(3+1+1)} = \frac{1}{5}.$$

Recall that for the same example we had $\text{SSI}(p_1) = \frac{2}{3}$, $\text{SSI}(p_2) = \frac{1}{6}$, and $\text{SSI}(p_3) = \frac{1}{6}$.

....

■ 3.5 TWO METHODS OF COMPUTING BANZHAF POWER

This section presents two new procedures for calculating total Banzhaf power. Both procedures begin with a very simple chart that has the winning coalitions enumerated in a vertical list down the left side of the page, and the individual voters enumerated in a horizontal list across the top. For example, if the yes–no voting system is the original European Economic Community, the chart (with "F" for "France" etc.) will have:

$$\textbf{F} \quad \textbf{G} \quad \textbf{I} \quad \textbf{B} \quad \textbf{N} \quad \textbf{L}$$

across the top. Down the left side it will have the fourteen winning coalitions which turn out to be (displayed horizontally at the moment for typographical reasons):

FGI, FGBN, FIBN, GIBN
FGIL, FGBNL, FIBNL, GIBNL
FGIB, FGIN
FGIBL, FGINL
FGIBN
FGIBNL

Notice the order in which we have chosen to list the winning coalitions: the first four are precisely the ones with weight 12, the next four are the ones with weight 13, then the two with weight 14, the two with weight 15, the one with weight 16, and the one with weight 17. If the voting system is weighted, this is a nice way to ensure that no winning coalitions have been missed. In what follows, we shall need the observation that there are fourteen winning coalitions in all.

We now present and illustrate the two procedures for calculating total Banzhaf power. Notice that "critical defection" is not mentioned in either procedure.

PROCEDURE 1. Assign each voter (country) a "plus one" for each winning coalition of which it *is* a member, and assign it a "minus one" for each winning coalition of which it *is not* a member. The sum of these "plus and minus ones" turns out to be the total Banzhaf power of the voter. (The reader wishing to get ahead of us should stop here and contemplate why this is so.) Continuing with the European Economic Community as an example, we have:

	F	G	I	B	N	L
FGI	1	1	1	−1	−1	−1
FGBN	1	1	−1	1	1	−1
FIBN	1	−1	1	1	1	−1
GIBN	−1	1	1	1	1	−1
FGIL	1	1	1	−1	−1	1
FGBNL	1	1	−1	1	1	1
FIBNL	1	−1	1	1	1	1
GIBNL	−1	1	1	1	1	1
FGIB	1	1	1	1	−1	−1
FGIN	1	1	1	−1	1	−1
FGIBL	1	1	1	1	−1	1
FGINL	1	1	1	−1	1	1
FGIBN	1	1	1	1	1	−1
FGIBNL	1	1	1	1	1	1
TBP(sum)	10	10	10	6	6	0

PROCEDURE 2. Assign each voter (country) a "plus two" for each winning coalition in which it appears (and assign it nothing for those in which it does not appear). Subtract the total number of winning coalitions from this sum. The answer turns out to be the total Banzhaf power of the voter. Continuing with the European Economic Community as an example, we have:

	F	G	I	B	N	L
FGI	2	2	2			
FGBN	2	2		2	2	
FIBN	2		2	2	2	
GIBN		2	2	2	2	
FGIL	2	2	2			2
FGBNL	2	2		2	2	2
FIBNL	2		2	2	2	2
GIBNL		2	2	2	2	2
FGIB	2	2	2	2		
FGIN	2	2	2		2	
FGIBL	2	2	2	2		2
FGINL	2	2	2		2	2
FGIBN	2	2	2	2	2	
FGIBNL	2	2	2	2	2	2
(sum)	24	24	24	20	20	14
Minus number of winning coalitions	−14	−14	−14	−14	−14	−14
TBP	10	10	10	6	6	0

The following chart summarizes the Banzhaf indices (arrived at by dividing each country's total Banzhaf power by $10+10+10+6+6+0 = 42$). This is analogous to what we did for the Shapley-Shubik indices in **Section 3.2**.

Country	Votes	Percentage of votes	BI	Percentage of power
France	4	23.5	5/21	23.8
Germany	4	23.5	5/21	23.8
Italy	4	23.5	5/21	23.8
Belgium	2	11.8	3/21	14.3
Netherlands	2	11.8	3/21	14.3
Luxembourg	1	5.9	0	0

Why is it that these two procedures give us the number of critical defections for each voter? Let's begin with the following easy observation: Procedure 2 yields the same numbers as does Procedure 1. That is, in going from Procedure 1 to Procedure 2, all the "minus ones" became "zeros" and all the "plus ones" became "twos." Hence, the sum for each voter increased by one for each winning coalition. Thus, when we subtracted off the number of winning coalitions, the result from Procedure 2 became the same as the result from Procedure 1.

So we need only explain why Procedure 1 gives us the number of critical defections that each voter has. (Recall that we are considering only monotone voting systems.) The key to understanding what is happening in Procedure 1 is to have at hand a particularly revealing enumeration of the winning coalitions. Such a revealing enumeration arises from focusing on a single voter p, with different voters in the role of p giving different enumerations. To illustrate such an enumeration, let's let the fixed voter p be the country Belgium in the European Economic Community. The list of winning coalitions corresponding to the fixed voter p will be made up of three "blocks" of coalitions:

Block 1: Those winning coalitions that do not contain p.

Block 2: The coalitions in Block 1 with p added to them.

Block 3: The rest of the winning coalitions.

For example, with Belgium playing the role of the fixed voter p, we would have the fourteen winning coalitions in the European Economic Community listed in the following order:

Block 1:	FGI
	FGIL
	FGIN
	FGINL
Block 2:	FGIB
	FGILB
	FGINB
	FGINLB
Block 3:	FGNB
	FINB
	GINB
	FGNLB
	FINLB
	GINLB

There are several things to notice about the blocks. First, the coalitions in Block 2 are all winning because those in Block 1 are winning and we are only considering monotone voting systems. Second, there are exactly as many coalitions in Block 2 as in Block 1, because if X and Y are two distinct winning coalitions in Block 1, and thus neither contains p, then adding p to each of X and Y will again result in distinct coalitions in Block 2. Moreover, every coalition in Block 2 arises from one in Block 1 in this way. Third, every coalition in Block 3 contains p, since all those not containing p were listed in Block 1.

Finally, and perhaps most importantly, is the observation that p's defection from a winning coalition is critical precisely for the coalitions in Block 3. That is, p does not even belong to the coalitions in Block 1, and p's defection from any coalition in Block 2 gives the corresponding winning coalition in Block 1, and thus is not critical. However, if p's defection from a coalition X in Block 3 were to yield a coalition Y that is winning, then Y would have occurred in Block 1, and so X would have occurred in Block 2 instead of Block 3.

The reason Procedure 1 works is now clear: The minus ones in Block 1 are exactly offset by the plus ones in Block 2, thus leaving a plus one contribution for each coalition in Block 3 and these are precisely the ones for which p's defection is critical.

Other consequences also follow. For example, a monotone yes-no voting system with exactly seventy-one winning coalitions has no dummies as defined in Exercise 18 in Chapter 2. (Exercise 19 asks why.) Notice that the listing of winning coalitions corresponding to the fixed voter p is used only to understand why the procedures work—such listings need not be constructed to actually calculate Banzhaf power using either Procedure 1 or Procedure 2.

Power indices tend to have some paradoxical aspects. For example, Felsenthal and Machover (1994) noticed the following paradoxical result for the Banzhaf index. Consider the weighted voting system in which there are five voters with weights 5, 3, 1, 1, 1 and the quota is 8. We denote this by:
$$[8 : 5, 3, 1, 1, 1].$$

The Banzhaf indicies of the voters turn out to be $\frac{9}{19}$, $\frac{7}{19}$, $\frac{1}{19}$, $\frac{1}{19}$, and $\frac{1}{19}$ (Exercise 32). Now suppose that the voter with weight 5 gives one of his "votes" to the voter with weight 3. This results in the weighted system

$$[8 : 4, 4, 1, 1, 1].$$

It now turns out (Exercise 32 again) that the first voter has Banzhaf index $\frac{1}{2}$. But $\frac{1}{2}$ is greater than $\frac{9}{19}$! (surprise, not factorial). Hence, by giving away a single vote to a single player (and no other changes being made), a player has increased his power as measured by the Banzhaf index. (Part of what is going on here is that the transfer of a vote from the first player to the second makes each of the last three players a dummy. Hence, the first two players together share a larger fraction of the power than they previously did, and—as one would expect—the second player gains more than the first. The trade-off is that the first player is gaining more from the gain caused by the effective demise of the last three voters than he is losing from the transfer of one vote from himself to the second voter.) More on this paradox is found in Exercise 33.

It turns out (as pointed out by Felsenthal and Machover) that the Shapley-Shubik index is not vulnerable to this particular type of paradox. But the Shapley–Shubik index is not immune to such quirks: Exercise 34 presents a paradoxical aspect (due to William Zwicker) of the Shapley–Shubik index that is not shared by the Banzhaf index.

■ 3.6 THE POWER OF THE PRESIDENT

Among the yes-no voting systems we have discussed, perhaps none is of more interest than the United States federal system. This brings us to an obvious question: What do the power indices have to say about the fraction of power held by the president in the U.S. federal system?

The calculations we will be doing (especially for the Shapley–Shubik index of the president) require some mathematical preliminaries. As a simple illustration of the first such preliminary we must confront, suppose we have four objects: $a, b, c,$ and d. In how many ways can we choose two of them (assuming that the order in which we choose them does not matter)? The answer turns out to be six: $ab, ac, ad, bc, bd,$ and cd. In general, if we start with n objects (instead of four) and ask for the number of ways we can choose k of them, where k is between 1 and n, then the following notation is used:

NOTATION. If $1 \leq k \leq n$, then the phrase "n choose k," denoted

$$\binom{n}{k},$$

refers to the number of distinct ways we can choose exactly k objects from a collection of exactly n objects.

The example above shows that "four choose two" equals six. The following proposition gives a relatively easy way to calculate these values.

PROPOSITION. *For $1 \leq k \leq n$ (and the convention that $0! = 1$), the following holds:*

$$\binom{n}{k} = \frac{n!}{k!(n-k)!}$$

PROOF. Recall from **Section 3.2** that the number of different ways we can arrange n objects is given by $n!$. With k fixed, we can think of each such arrangement of the n objects as being obtained by the following three-step process:

1. Choose k of the objects to be the initial "block."

2. Arrange these k objects in some order.

3. Arrange the remaining $n - k$ objects in some order.

For example, if the objects are a, b, c, d, e, and f, and $k = 3$, then step 1 might consist of choosing a, d, and f. Step 2 might consist of choosing the following arrangement of the three chosen objects: f followed by a followed by d. Step 3 might consist of choosing the following arrangement of the remaining objects: e follwed by b followed by c. These three steps yield the arrangement: $f \, a \, d \, e \, b \, c$.

Step 1 above can be done in n choose k different ways. Step 2 can be done in $k!$ different ways. Step 3 can be done in $(n - k)!$ different ways. Hence, according to the general multiplication principle from

Section 3.2, the number of different ways the three-step process can be done is arrived at by multiplying these three numbers together. That is, the number of distinct arrangements of the n objects arrived at by the three step process is:

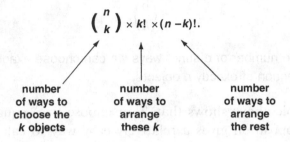

$$\binom{n}{k} \times k! \times (n-k)!.$$

number of ways to choose the k objects

number of ways to arrange these k

number of ways to arrange the rest

Moreover, it should be clear that every arrangement of the n objects can be uniquely arrived at by the above three-step process, and we already know there are $n!$ such arrangements. Thus,

$$\binom{n}{k} \times k! \times (n-k)! = n!$$

and so, dividing both sides by $k! \times (n-k)!$ yields

$$\binom{n}{k} = \frac{n!}{k!(n-k)!}$$

as desired. This completes the proof.

With these preliminaries at hand, we can now turn to the task of calculating the power of the president according to the two different power indices that have been introduced. In the version of the U.S. federal system we will consider, the tie-breaking role of the vice president is ignored. We consider each of the two power indices in turn.

The Shapley-Shubik Index of the President

Using the n choose k notation, it takes only a few lines to write down the arithmetic expression giving the Shapley-Shubik index of the president (and the reader who wishes to see it now can flip a few pages ahead and find it there). However, explaining where this expression came from is quite another story (and the one we want to tell). So, let's consider a simpler version of the federal system—a "mini-federal system"—where there are only six senators and six members of the House and the president. (We choose the number six because it is the smallest positive integer for which half it and two-thirds it are also integers, and these are the fractions involved in the rules for passage.) Passage in the mini-federal system requires either two-thirds of both houses or half of each house and the president.

For the president to be pivotal in an ordering of the thirteen voters in our mini-federal system, he must be preceded by at least three members of the House and at least three members of the Senate, but by fewer than four members of at least one of the two chambers. This can happen in the following seven ways:

1. Three House members and three senators precede the president in the ordering (and, thus, three House members and three senators follow the president in the ordering).

2. Three House members and four senators precede the president in the ordering (and, thus, three House members and two senators follow the president in the ordering).

3. Three House members and five senators precede the president in the ordering (and, thus, three House members and one senator follow the president in the ordering).

4. Three House members and six senators precede the president in the ordering (and, thus, three House members and no senators follow the president in the ordering).

5. Four House members and three senators precede the president in the ordering (and, thus, two House members and three senators follow the president in the ordering).

6. Five House members and three senators precede the president in the ordering (and, thus, one House member and three senators follow the president in the ordering).

7. Six House members and three senators precede the president in the ordering (and, thus, no House members and three senators follow the president in the ordering).

We wish to count how many orderings of each of the seven kinds there are. Consider first the orderings in the first entry on the list. Each such ordering can be built in a four-step process:

Step 1: Choose three of the six House members to precede the president in the ordering. This can be done in six choose three ways.

Step 2: Choose three of the six senators to precede the president in the ordering. This can be done in six choose three ways.

Step 3: Choose an ordering of the six people from steps 1 and 2 who will precede the president. This can be done in 6! ways.

Step 4: Choose an ordering of the six people (the remaining House members and senators) who will come after the president. This can be done in 6! ways.

By the multiplication principle, we know that the total number of orderings that can be constructed by the above four-step process is:

$$\binom{6}{3}\binom{6}{3}6!6!.$$

A similar argument yields a similar expression for the number of orderings that arise in the other six entries on the list. The sum of these seven expressions gives us the total number of orderings of the thirteen voters for which the president is pivotal. Hence, to obtain the Shapley-Shubik index of the president in this mini-federal system, we simply divide that result by 13! This yields:

$$\frac{\binom{6}{3}\binom{6}{3}6!6! + 2\binom{6}{3}\binom{6}{5}7!5! + 2\binom{6}{3}\binom{6}{5}8!4! + 2\binom{6}{3}\binom{6}{6}9!3!}{13!}.$$

This evaluation can be done by hand, and we leave it for the reader.

The following expression gives the numerator for the Shapley–Shubik index of the president in the U.S. federal system (with the vice president ignored). The denominator is 536 factorial. We leave it to the reader (see Exercise 39) to provide an explanation for this expression that is analogous to what we did for the mini-federal system.

$$\binom{435}{218}\left[\binom{100}{51}(218+51)!(535-218-51)!+\cdots\right.$$

$$\left.+\binom{100}{100}(218+100)!(535-218-100)!\right]$$

$$+\cdots$$

$$+\binom{435}{289}\left[\binom{100}{51}(289+51)!(535-289-51)!+\cdots\right.$$

$$\left.+\binom{100}{100}(289+100)!(535-289-100)!\right]$$

$$+\binom{435}{290}\left[\binom{100}{51}(290+51)!(535-290-51)!+\cdots\right.$$

$$\left.+\binom{100}{66}(290+66)!(535-290-66)!\right]$$

$$+\cdots$$

$$+\binom{435}{435}\left[\binom{100}{51}(435+51)!(535-435-51)!+\cdots\right.$$

$$\left.+\binom{100}{60}(435+66)!(435-218-66)!\right]$$

One would not want to simplify such an expression by hand. Fortunately, there are computer programs available—like *Mathematica*—which make things easy. For example, to evaluate the above expression (including the division by 536!) one simply types the following as input for *Mathematica*:

Sum[Binomial [435, h] Binomial[100, s]$(s+h)!(535-s-h)!$,

$\{h, 218, 289\}, \{s, 51, 100\}$]/536!+

Sum[Binomial [435, h] Binomial[100, s]$(s+h)!(535-s-h)!$,

$\{h, 290, 435\}, \{s, 51, 66\}$]/536!

One then simply waits (how long depends upon how fast your computer is) until *Mathematica* responds with:

120596538268818663439104326960166\2

11644652437238111390576\

175712831370645194187827110103 6003/

7515229940063793084403227234743776

3002488117950021866418 4\

4565883705084200048715438500735040

Finding that response a little unsettling, one types "N[%]". This instructs *Mathematica* to express the answer as a nice decimal. The output is then

0.16047.

Thus, according to the Shapley–Shubik index, the president has about 16 percent of the power in the U.S. federal system.

The Banzhaf Index of the President

The Banzhaf index of power of the president is obtained by dividing his total Banzhaf power by the sum of the Banzhaf powers of all the voters in the U.S. federal system (i.e., the president, the 100 members of the Senate, and the 435 members of the House – for simplicity, we are still ignoring the vice president). Thus, to calculate the Banzhaf index of the president, we need to determine not only his total Banzhaf power, but also that of each member of the Senate and each member of the House.

We will make these calculations of total Banzhaf power by using the second procedure in **Section 3.5**, wherein a voter's total Banzhaf power was shown to be simply twice the number of winning coalitions to which that voter belongs minus the total number of winning coalitions. A little notation will make things easier.

Let S denote the number of coalitions within the Senate that contain at least two-thirds of the members of the Senate. Thus,

$$S = \binom{100}{67} + \cdots + \binom{100}{100}.$$

Let S_p denote the number of coalitions within the Senate that contain at least two-thirds of the members of the Senate and that contain a particular senator p. Such a coalition is arrived at by choosing anywhere from 66 to 99 of the other senators. Thus,

$$S_p = \binom{99}{66} + \cdots + \binom{99}{99}.$$

Let s denote the number of coalitions within the Senate that contain at least one-half of the members of the Senate. Thus,

$$s = \binom{100}{50} + \cdots + \binom{100}{100}.$$

Let s_p denote the number of coalitions within the Senate that contain at least one-half of the members of the Senate and that contain a particular senator p. Such a coalition is arrived at by choosing anywhere from 49 to 99 of the other senators. Thus,

$$s_p = \binom{99}{49} + \cdots + \binom{99}{99}.$$

Let H denote the number of coalitions within the House that contain at least two-thirds of the members of the House. Thus,

$$H = \binom{435}{290} + \cdots + \binom{435}{435}.$$

Let H_p denote the number of coalitions within the House that contain at least two-thirds of the members of the House and that contain a particular member of the House p. Such a coalition is arrived at by choosing anywhere from 289 to 434 of the other members of the House. Thus,

$$H_p = \binom{434}{289} + \cdots + \binom{434}{434}.$$

Let h denote the number of coalitions within the House that contain at least one-half of the members of the House. Thus,

$$h = \binom{435}{218} + \cdots + \binom{435}{435}.$$

Let h_p denote the number of coalitions within the House that contain at least one-half of the members of the House and that contain a particular member of the House p. Such a coalition is arrived at by choosing anywhere from 217 to 434 of the other members of the House. Thus,

$$h_p = \binom{434}{217} + \cdots + \binom{434}{434}.$$

It is now easy to write down expressions involving $S, S_p, s, s_p, H, H_p,$ h_1 and h_p (and the n choose k notation) that give us the total Banzhaf power for the president, a member of the House, and a member of the Senate. (Recall that the desired expression is simply two times the number of winning coalitions to which a voter belongs, with the total number of winning coalitions then subtracted from this.) We leave this for the reader. However, as was the case for the Shapley–Shubik index, actual calculations need be done on a computer. The results turn out to be:

BI(the president)   = .038.
BI(a senator)   = .0033.
BI(a member of the House) = .0015.

A more meaningful way to view these results is in terms of percentages (of power) as opposed to small decimals. It is also more meaningful to consider the power of the Senate as opposed to the power of a single senator, and to do the same for the House (assuming that power is additive—a risky assumption at best). The results then become

(Banzhaf) Power held by the president = 4%
(Banzhaf) Power held by the Senate   = 33%
(Banzhaf) Power held by the House   = 63%

■ 3.7 THE CHAIR'S PARADOX

Our previous considerations of political power have focused on quantitative measures of influence over outcomes. In this section, we change gears slightly to give quite another view of this somewhat elusive concept of power. We present a classic result known as the Chair's paradox.

The primary purpose of introducing this paradox is to illustrate the extent to which apparent power need not correspond to control over outcomes.

The situation we want to consider is the following. There are three people, A, B, and C, and three alternatives, a, b, and c. The preference lists of the three people are as illustrated below (and replicate those from the voting paradox of Condorcet in Chapter 1).

A	B	C
a	b	c
b	c	a
c	a	b

The social choice procedure being used is somewhat different from those we have considered before. That is, the preference lists will not be regarded as inputs for the procedure, but will only be used to "test" the extent to which each of A, B, and C should be happy with the social choice. The social choice will be determined by a standard voting procedure where voter A (the Chair) also has a tie-breaking vote. The point of not using the preference lists themselves is that we do not want to force any one of the three players to vote for his or her top choice, although it is probably not clear at the moment why anything else would benefit any of them. In fact, voting where everyone is bound to vote for his or her top choice is called *sincere voting*. Anything else (and this is our interest) is called *sophisticated voting*.

In our present situation, a *strategy* for any of the three people A, B, or C is simply a choice for which of the three alternatives a, b, c to vote. A sequence of such votes will be called a *scenario*.

DEFINITION. Fix a player P and consider two strategies $V(x)$ and $V(y)$ for P. [Think of $V(x)$ as being "vote for alternative x."] Then we'll say that $V(x)$ weakly dominates $V(y)$ for player P provided that the following both hold:

1. For every possible scenario (i.e., choice of alternatives for which to vote by the other players), the social choice resulting from $V(x)$ is at least as good for Player P (as measured by his or her preference list) as that resulting from $V(y)$.

2. There is at least one scenario in which the social choice resulting from $V(x)$ is strictly better for Player P than that resulting from $V(y)$.

A strategy is said to be *weakly dominant for Player P* if it weakly dominates every other available strategy.

In determining whether or not a strategy is a weakly dominant one, we must consider all possible scenarios and compare the result achieved by using it with that achieved by using (in our case) either of the other two. What is needed, then, is a notation that will lay all of this out before us so that we can check for the desired preferences. In general, we'll do this by displaying the scenarios of votes by other players as a tree, and then listing the various strategies available to the player under consideration as a vertical column to the left, with the proposed dominant strategy at the bottom. Corresponding outcomes will then be exhibited. All of this is illustrated in the following three propositions, which refer, of course, to the setup described above.

PROPOSITION. *"Vote for alternative a" is a weakly dominant strategy for the Chairman A.*

PROOF. There are nine scenarios that arise from what players B and C might do, since each could vote for a, b, or c. These scenarios are presented in what is called a *tree* in Figure 5.

The "start" node on top has no significance except to hold the tree together. The next level down gives the three possibilities for B's vote. Below each of these, we have the same three possibilities for C to choose among. Thus, the "tree of scenarios" for players other than the Chair A is the top half of the following diagram. The columns below these nine scenarios give the outcomes (i.e., the social choices) depending upon whether Player A votes for c, b, or a (with the outcomes from top to bottom, respectively). For example, the column enclosed in a rectangle corresponds to the scenario where Player B votes for a and Player C votes for b. Now, if the Chair A votes for c, then it's a tie with one vote for each alternative, and A's tie-breaking vote now goes into effect to yield c as the outcome. This is the "c" at the top of the column. If A votes for b instead, then b wins by a two-to-one margin over a with c getting no votes. This is the "b" in the middle of the column. Finally if

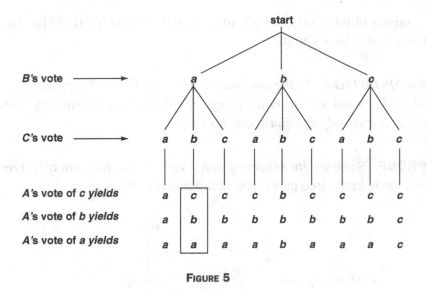

FIGURE 5

A votes for *a*, then it is the social choice with two votes, and this is the "*a*" at the bottom of the column.

With the scenarios and outcomes set out before us in this way, it is now easy to refer to player *A*'s preference list (*a* preferred to *b* preferred to *c*), and see that "vote for *a*" is, indeed, a weakly dominant strategy for *A*. That is, simply notice that the outcome at the bottom of each column is never worse for *A* than either of the outcomes above it, and that it is strictly better than both in at least one case (namely, the one enclosed in a rectangle). We should point out that it wasn't necessary to find a single scenario that simultaneously demonstrated the potential strict superiority of voting for *a* over voting for *b* and *c*. That is, we could have said that the second column shows voting for *a* can be better for *A* than voting for *b*, and the third column shows that voting for *a* can be better for *A* than voting for *c*. This completes the proof.

It turns out that neither Player *B* nor Player *C* has a weakly dominant strategy in the sense of the above proposition (see Exercise 40).

Thus, all that we can reasonably infer from what we have so far is that Player *A* appears to have no rational justification for voting for anything except alternative *a*. However, if we now assume that *A* will definitely go with his or her weakly dominant strategy of voting for *a*, then it is (literally) a whole new "game" and we can press on with an analysis of what rational self-interest will dictate for the other two players in this new situation.

> **PROPOSITION.** *In the new situation where Player A definitely uses his or her weakly dominant strategy of voting for a, the strategy "vote for c" is a weakly dominant one for Player C.*

PROOF. Since we are assuming that A votes for a, there are only three scenarios in the tree part of the notational presentation.

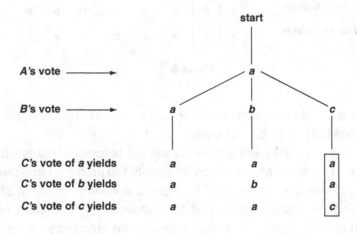

Column three shows there is a scenario (namely, B votes for c) where voting for c yields a strictly better result for C than voting for either b or a (since C prefers c to a). Notice again that we could have used column two to show that the strategy "vote for c" dominates "vote for b" and then used column three to show that "vote for c" dominates "vote for a." Of course, we also have to check that voting for c is at least as good for C as voting for a or b in every other scenario, but this amounts to just observing that C prefers a to b and c to a. This completes the proof.

Given the two previous propositions (and the common knowledge assumption that is implicitly being made), we have that rational self-interest yields a vote of a by A and a vote of c by C. This is not too surprising since both are simply voting for their top choice. The following, then, is somewhat more striking.

PROPOSITION. *In the new situation where Player A definitely votes for a and Player C definitely votes for c, the strategy "vote for c" is a weakly dominant one for Player B.*

PROOF. The desired notational presentation is as follows:

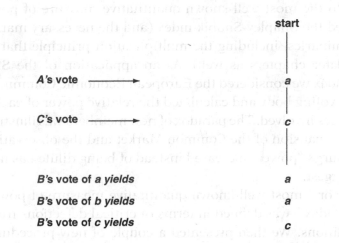

The proof is completed by simply observing that *B* prefers *c* to *a*.

Thus, the result of sophisticated voting (where everyone acts according to rational self-interest and knows that everyone else is doing the same, etc.) is:

A votes for *a*

B votes for *c*

C votes for *c*,

yielding an outcome of *c*. What is striking here is that *c* is *A*'s least preferred alternative even though *A* was the Chair and had the additional "power" provided by a tie-breaking vote in addition to his or her regular vote. It is also important to note that this outcome does not depend on any cooperative effort on the part of *B* and *C*; each is operating independently in his or her own self-interest without relying on the other doing anything except also acting in his or her own self-interest.

■ 3.8 CONCLUSIONS

We began this chapter with a hypothetical example illustrating that "fraction of votes" and "fraction of power" need not be the same. Turning to the most well-known quantitative measure of power, we introduced the Shapley-Shubik index (and the necessary mathematical preliminaries, including the multiplication principle that will be used in later chapters as well). As an application of the Shapley–Shubik index we considered the European Economic Community as a weighted voting body and calculated the relative power of each of the six countries involved. The paradox of new members was illustrated by the 1973 expansion of the Common Market and the observation that Luxembourg's "power" increased instead of being diluted as intuition would suggest.

The second most well-known quantitative measure of power—the Banzhaf index—was defined in terms of critical defections from winning coalitions. We then presented a couple of new procedures that allow one to calculate total Banzhaf power by making a single "run" down the list of winning coalitions, and we illustrated this with the European Economic Community.

With these two power indices at hand, we turned to the task of calculating the power of the president (as well as the House and Senate) in the U.S. federal system according to the two power indices introduced so far. This required introducing the "n choose k" notation and the proposition that allows us to calculate n choose k in terms of factorials. The power indices give somewhat different results. For example, the Banzhaf index suggests that the president has 4 percent of the power and the House holds roughly twice as much power as the Senate. The Shapley-Shubik index gave the president 16 percent of the power. So which of the two (or three or four) is more accurate, and how can we test this against actual experience with the federal government? Certainly, this is the right question to ask, but not of non-political scientists. Hence, we will content ourselves here with referring the reader to Brams, Affuso, and Kilgour (1989) and Packel (1981).

We concluded with the so-called "Chair's paradox," which presents a situation wherein individual preferences are such that even though the Chair has more "power" in terms of the voting scheme, he winds up with his least-preferred alternative when the other two act independently (and this independence of action is important to note) in their own best interests. The actual preference lists occurring here are the same ones that led to the voting paradox in Chapter 1.

EXERCISES

1. List out the 24 orderings of p_1, p_2, p_3, p_4. Arrange them so that the first four orderings in your list arise from the first orderings presented in Figure 3 in **Section 3.2** (i.e., $p_3 p_2 p_1$), the next four from the second ordering presented in Figure 3 in **Section 3.2** (i.e., $p_2 p_3 p_1$), etc.

2. Illustrate the twenty-four possible orderings of p_1, p_2, p_3, p_4 by drawing a tree with a start node at the top, four nodes on level one corresponding to a choice of which p_i will go first in the ordering, three nodes immediately below each of these on the next level corresponding to a choice of which p_i then will go second, etc.

3. Suppose we want to form a large governmental committee by choosing one of the two senators from each of the fifty states. How many distinct such committees can be formed? (Hint: Step 1 is to choose one of the two senators from Maine; there are two ways to do this. Step 2 is to choose ...) Comment on the size of this number.

4. Show that there are fewer than 362,881 different games of tic-tac-toe with "X" going first. Note that two games of tic-tac-toe are different if there is a number n, necessarily between one and nine, so that the symbol being played (i.e., "X" or "O") at move n is placed in different squares in the two games.

5. Show that there are fewer than 20,000 three-letter words in the English language.

6. If your college has 35 different academic departments, and each department offers, on average, 20 courses each semester, and you take four courses each semester, what is the total number of distinct sets of courses you have to choose from? Express your answer in n choose k notation.

7. Using n choose k and factorial notation, indicate the number of orderings in which the president is preceded by exactly 23 senators and exactly 235 members of the House.

8. Suppose that x has five votes, y has three votes, z has three votes, and w has two votes. Assume that eight votes are needed for passage. Calculate $SSI(x)$, $SSI(y)$, $SSI(z)$, and $SSI(w)$. Show all your work.

9. Show that Luxembourg has a Shapley-Shubik index of zero.

10. Show that the Netherlands and Belgium both have a Shapley-Shubik index of $9/60$ in two different ways:
    (a) By directly calculating it as we did for France.
    (b) By using the fact that the sum of the indices must be one.

11. In the calculation of the Shapley-Shubik index of France (within the European Economic Community), the details were increasingly omitted as we proceeded from case 1 through case 4. Redo case 4 in as much detail as was provided for case 1.

12. Consider the 1973 expansion of the European Economic Community as described in **Section 3.3**. Show that Luxembourg has positive Shapley–Shubik index by producing an ordering of the countries involved for which Luxembourg is pivotal.

13. Show that even if Luxembourg had been left with just one vote in the expansion of 1973 (with everything else as it actually was in the expansion), the Shapley-Shubik index of Luxembourg still would have been nonzero.

14. In the 1973 expansion of the European Economic Community, the percentage of votes needed for passage rose the trivial amount from 70.6 percent to 70.7 percent. Suppose they had decided to require forty (instead of forty-one) votes for passage. Would a paradox of new members still have taken place?

15. Consider a voting system for the six New England states where there are a total of seventeen votes and twelve or more are required for passage. Votes are distributed as follows:

$$\text{MA:4} \quad \text{ME:3} \quad \text{NH:2}$$
$$\text{CT:4} \quad \text{RI:3} \quad \text{VT:1}$$

    (a) Calculate SSI(MA).
    (b) Calculate SSI(ME).

(c) Calculate SSI(NH).

(d) Calculate SSI(VT).

16. Suppose that x has five votes, y has three votes, z has three votes, and w has two votes. Assume that eight votes are needed for passage.

   (a) Calculate TBP for each voter by directly using the definition of TBP in terms of critical defections. Show all your work.

   (b) Calculate TBP for each voter by using Procedure 1 from **Section 3.4**. (There are seven winning coalitions.)

   (c) Calculate TBP for each voter by using Procedure 2 from **Section 3.4**.

17. Calculate total Banzhaf power and the Banzhaf indices for the New England states in the yes-no voting system from Exercise 15.

18. Explain how we know, in a monotone yes-no voting system, that every voter belongs to at least half the winning coalitions. Explain also why a voter in such a system is a dummy if and only if he belongs to exactly half the winning coalitions.

19. Explain how we know that there are no dummies in a monotone yes-no voting system with exactly 71 winning coalitions.

20. Suppose we have a yes-no voting system with four voters A, B, C, and D, where the winning coalitions are as follows:

$$ABCD, ABC, ABD, ACD, BCD, AB, AD$$

   (a) Compute the Banzhaf index of each voter.

   (b) Compute the Shapley-Shubik index of each voter.

21. Listed below are the number of electoral votes of each state for the 2004 presidential election.

States	Electoral Votes	States	Electoral Votes
CA	55	TX	34
NY	31	FL	27
IL, PA	21	OH	20
MI	17	GA, NJ, NC	15
VA	13	MA	12
IN, MO, TN, WA	11	AZ, MD, MN, WI	10
AL, CO, LA	9	KY, SC	8
CT, IA, OK, OR	7	AR, KS, MS	6
NE, NV, NM, UT, WV	5	HI, ID, ME, NH, RI	4
AK, DE, DC, MT, ND, SD, VT, WY	3		

Suppose that the New England states (ME, MA, RI, CT, VT, NH) decide to vote amongst themselves on a particular issue. They use their electoral votes above to create a six-state weighted voting system.

(a) Calculate the Banzhaf index of each state.

(b) Calculate the Shaley-Shubik index of each state.

22. Suppose that we have a yes-no voting system with three voters A, B, and C, where the winning coalitions are as follows:

$$ABC, AB, AC, A$$

We saw in Exercise 22 in Chapter 2 that A is a dictator in this system.

(a) Calculate the Banzhaf index of each of the three voters.

(b) Calculate the Shapley-Shubik index of each of the three voters.

23. Consider the weighted yes-no voting system in which four voters A, B, C, and D have 8, 8, 4, and 2 votes, respectively and the quota is 14.

(a) Calculate the Banzhaf index of each of the four voters.

(b) Calculate the Shapley-Shubik index of each of the four voters.

24. Consider the mini-federal system with thirteen votes wherein there are six House members, six senators, and the president, and passage requires half the House, half the Senate, and the president, or two-thirds of both houses.

(a) Calculate the Banzhaf index of the president.

(b) Calculate the Shapley-Shubik index of the president.

25. Assume that we have a four-person weighted voting system in which one voter has 3 votes, two voters have 2 votes, and one voter has 1 vote and the quota is 5. Calculate the Banzhaf index of each voter.

26. Consider a 3-person yes-no voting system with voters A, B, and C. Suppose that a coalition is winning if it contains either A or both B and C.

(a) Calculate the Shapley-Shubik Index of each voter.

(b) Calculate the Banzhaf Index of each voter.

27. Assume that we change the quota in the EEC from 12 to 11, leaving the weights of the countries as they were. Calculate SSI(Luxembourg) in this new system.

28. Consider the minority veto system where there are six voters: A, B, C, D, E, and F, two of whom (E and F) form a designated minority,

and passage requires a total of at least four of the six votes and at least one of the minority votes. Calculate the Banzhaf index for each of the voters in this system. (There are 21 winning coalitions.)

29. Suppose we have a minority veto yes-no voting system with five voters: majority voters A, B, C and minority voters R and S. Passage requires a total of at least three votes, at least one of which must be a minority voter.
    (a) Write down all winning coalitions for this yes-no voting system.
    (b) Calculate the Banzhaf index of each voter.

30. Amanda, Ben, Chris, Desdemona, and Ed spend a lot of time together. They've created their own yes-no voting system to help decide what to do each Saturday. Amanda has 5 votes since she's the glue that holds everyone together. Ben and Cate each have 4 votes, and Desdemona and Ed each get 3 votes because they are newer to the group and sometimes a little annoying. Calculate the Shapley-Shubik index of Desdemona.

31. Suppose that A is a dummy voter in a yes-no voting system.
    (a) Does the Banzhaf index of A have to be 0? Prove that this is true or give an example of a system where this is false.
    (b) Does the Shapley Shubik index of A have to be 0? Prove that this is true or give an example of a system where this is false.

32. Use Procedure 1 or 2 to verify the calculations in the Felsenthal-Machover example in **Section 3.5**.

33. (a) Discuss whether or not you find it paradoxical that in going from the system $[8 : 5, 3, 1, 1, 1]$ to the system $[8 : 4, 4, 0, 0, 0]$, the first player's Banzhaf power increases.
    (b) Discuss whether or not you find it paradoxical that two different choices of weights, such as $[8 : 4, 7, 0, 0, 0]$ and $[8 : 4, 4, 1, 1, 1]$ give the same yes-no voting system.
    (c) What, if anything, do your responses to (a) and (b) say about the Felsenthal–Machover paradox?

34. The Shapley–Shubik index is not without paradoxical aspects. For example, the following was pointed out by William Zwicker. Suppose we have a bicameral yes–no voting system wherein an issue must win in both the House and the Senate in order to pass. (We are not assuming that the House and Senate necessarily use majority rule, but we are assuming they have no common members.) Suppose that both you and Fred belong to the House and that—when the

House is considered as a yes–no voting system in its own right—Fred has three times as much "power" as you have. Then shouldn't Fred still have three times as much power as you have when we consider the bicameral yes-no voting system? (This is a rhetorical question.)

(a) Suppose that $X_1, \ldots, X_m$ are the winning coalitions in the House and $Y_1, \ldots, Y_n$ are the winning coalitions in the Senate. Assume that Fred belongs to t of the winning coalitions in the House and that you belong to z of the winning coalitions in the House.

   1. Show that there are mn winning coalitions in the bicameral system.

   2. Use Procedure 2 to show that Fred's total Banzhaf power in the House is $2t - m$ and that yours is $2z - m$.

   3. Use Procedure 2 to show that Fred's total Banzhaf power in the bicameral system is $2tn - mn$ and that yours is $2zn - mn$.

   4. Show that if Fred has v times as much power (as measured by the Banzhaf index) in the House as you have, then Fred also has v times as much power (as measured by the Banzhaf index) in the bicameral system as you have.

(b) Suppose the House consists of you, Fred, and Bill, and suppose there are two minimal winning coalitions in the House: Fred alone (as one), and you and Bill together (as the other). In the Senate, there are two people and each alone is a minimal winning coalition.

   1. Show that in the House alone, Fred has four times as much power—as measured by the Shapley-Shubik index—as you have. (The values turn out to be $\frac{4}{6}, \frac{1}{6}, \frac{1}{6}$.)

   2. Show that in the bicameral system, this is no longer ture. (The values turn out to be $\frac{44}{120}$ for me and $\frac{14}{120}$ for you.)

35. Calculate each of the following:
   (a) five choose three,
   (b) four choose four,
   (c) six choose two, and
   (d) six choose four.

36. How many three-member subcommittees can be formed from a parent committee of size nine?

37. (a) List out the distinct ways of choosing two objects from the five objects *a, b, c, d, e*.
    (b) List out the distinct ways of choosing three objects from the five objects *a, b, c, d, e*.
    (c) Without mentioning the "factorial formula" from this section, explain why the number of pairs you found in (a) was the same as the number of triples you found in (b).

38. One way to obtain an ordering of the 536 voters in the United States federal system for which the president is pivotal is to have him preceded in the order by (say) 250 members of the House and by (say) 60 senators. Using the "*n* choose *k*" and "factorial" notation, write down an expression that gives the total number of such orders.

39. The first two lines in the expression used to get the Shapley–Shubik index of the president was:

$$\binom{435}{218}\left[\binom{100}{51}(218+51)!(535-218-51)! + \ldots\right.$$
$$\left. + \binom{100}{100}(218+100)!(535-218-100)!\right]$$

This represents a collection of orderings for which the president is pivotal. Describe this collection of orderings.

40. In the Chair's paradox, show that neither *B* nor *C* has a weakly dominant strategy in the sense of the first proposition in **Section 3.6**.

41. Suppose that in the Chair's paradox situation, person *C* changes his preferences to *b* over *a* over *c*. Is "vote for *a*" still a weakly dominant strategy for the Chair *A*?

42. Suppose that we have three people I, II, III and three alternatives *a, b, c*. Suppose the preference orderings are as follows:

I	II	III
a	a	b
b	c	a
c	b	c

Consider the following semi-strange social choice procedure: Alternative *c* wins unless two or more people vote for *a* (in which case *a* wins) or all three people vote for *b* (in which case *b* wins). Show that "vote for *b*" weakly dominates "vote for *c*" for Player I.

CHAPTER
4

Conflict

··

■ 4.1 INTRODUCTION

One of the central concepts of political science is conflict, that is, situations where the actions of one individual (or group) both influence and are influenced by those of another. Real-world examples of such conflict situations tend to be enormously complex, and a considerable amount of influential work in political science deals with the analysis of particular conflict situations and the ramifications of literally dozens of subtle influences upon the events that took place.

The kind of analysis that we will undertake here, however, is at the other end of the spectrum. Instead of the kind of fine analysis that is very specific to a particular event, we will consider some extremely simple game-theoretic models that provide a very coarse analysis applicable to many different events of historical significance. The justification for this undertaking, however, lies in the extent to which these game-theoretic analyses, coarse though they may be, nevertheless shed light on why various events unfolded as they did, as well as to explain some of the intractabilities of situations such as the arms race of the 1960s, 1970s, and 1980s.

We begin in **Section 4.2** by introducing "2 × 2 games" (read as "two-by-two games"), so called because they involve two parties each of which is choosing one of two available strategies. In **Section 4.3** we introduce the notion of dominant strategies and Nash equilibria in this context. **Sections 4.4, 4.5,** and **4.6** then examine 2 × 2 games that model three real-world situations: the arms race, the Cuban missile crisis, and the Yom Kippur War. The first model works extremely well, the second moderately well, and the third model fails miserably. Yet, in **Section 4.7**, we show how to embellish the model of the Yom Kippur War via something called "the theory of moves," and this, indeed, gives a most satisfying analysis.

■ 4.2 TWO-BY-TWO GAMES

The games that our models will be based on are called "2 × 2 ordinal games." The framework for such a game is as follows:

1. There are two players: Row and Column.

2. Each player has a choice of two alternatives: C (for "cooperate") or N (for "noncooperate"). A choice of an alternative is called a *strategy*.

3. The play of the game consists of a single move: Row and Column simultaneously (and independently) choose one of the two alternatives, C or N. This yields four possible outcomes as displayed in Figure 1.

4. Each player ranks the four possible outcomes according to his or her relative preference. The outcome considered "best" (by, say, Row) is labeled "4" (by Row); second best, "3"; third, "2"; and, finally, the outcome considered worst (still, by Row) is labeled "1" (by Row).

These games are called "ordinal" games since the labels 4, 3, 2, and 1 for the outcomes reflect only the order of preference as opposed to the (absolute) magnitude of one's preference for any particular outcome. Thus, for example, an outcome (say CN) labeled "4" by Row should not

	Row's Choice	Column's Choice	Shorthand Notation
Outcome a	C	C	CC
Outcome b	C	N	CN
Outcome c	N	C	NC
Outcome d	N	N	NN

FIGURE 1

			Column	
			C	N
Row's Preference Ranking		C	3	1
	Row			
		N	4	2

FIGURE 2

			Column	
			C	N
Column's Preference Ranking		C	3	4
	Row			
		N	1	2

FIGURE 3

be construed as twice as good (in Row's view) as an outcome labeled "2" by Row.

For the sake of illustrating the notation we will use, let us look at a particular 2 × 2 ordinal game—one that will, in fact, turn out to be important. Describing a 2 × 2 ordinal game means specifying a total of eight things: Row's preference ranking of the four possible outcomes CC, CN, NC, NN, and Column's ranking of the same four possible outcomes. For the particular example we want to use here, the preference rankings are shown in Figures 2 and 3 above.

Thus, Row ranks the four outcomes, from best to worst, as NC, CC, NN, CN, and Column ranks the outcomes, from best to worst, as CN, CC, NN, NC.

The rectangular arrays we have used to describe Row and Column's preferences correspond to mathematical objects called "matrices" (plural of "matrix"), more explicitly, "2×2 matrices," since each array has two rows (i.e., two horizontal sequences of numerical entries) and two columns (i.e., two vertical sequences of numerical entries). This explains the choice of "Row" and "Column" as names for the players. Notice also that in the 2×2 game described above, both Row and Column prefer the outcome CC to the outcome NN. That is, both assign mutual cooperation (CC) a "3" (second best) and mutual noncooperation a "2" (second worst). In particular, a gain for one player is not necessarily a loss for the other. For this reason these games are called "variable-sum" as opposed to "zero-sum."

The standard notation for presenting a particular 2×2 ordinal game involves using a single 2×2 matrix to simultaneously present the preference rankings of Row and Column. Each of the four entries in this case involves two numbers: Row's ranking and Column's ranking. Thus, for example, if we consider the upper right hand entry (that is, the one that is simultaneously in the first row and the second column), we find that Row ranks it, in our example, as "1" and Column ranks it as "4." Hence in the single matrix display of both Row's preferences and Column's preferences, we could use something like "1/4" or "(1, 4)" as the upper right hand entry as long as we agree that the first number so displayed applies to Row and the second number to Column. We'll opt for the "ordered pair" notation (1,4). Thus, the single 2×2 matrix representing the game described above is shown in Figure 4 below.

		Column	
		C	N
Row	C	(3, 3)	(1, 4)
	N	(4, 1)	(2, 2)

FIGURE 4

There aren't all that many 2 × 2 ordinal games, and most of those are (and probably will remain) both uninteresting and unimportant. In the rest of this chapter, however, we will concentrate on what are the most well known and probably most interesting of the lot: Prisoner's Dilemma and Chicken.

■ 4.3 DOMINANT STRATEGIES AND NASH EQUILIBRIA

Recall that a strategy (for, say, Row) in a 2 × 2 ordinal game is a choice of C or N. Recall also that an outcome in a 2 × 2 ordinal game is an ordered pair, and that, for example, the outcome (3, 1) would be preferred by Row to the outcome (2, 4). For brevity, we might simply say that (3, 1) is better for Row than (2, 4). The central idea of this section is the following.

DEFINITION. The strategy N is said to be *dominant for Row* in a (particular) 2 × 2 ordinal game if—regardless of what Column does—it yields an outcome that is better for Row than would have been obtained by Row's use of the strategy C.

We could, of course, similarly define the notions of *C being dominant for Row, N being dominant for Column*, and *C being dominant for Column*. Illustrations of these concepts will occur in **Sections 4.4** and **4.5** where we consider, respectively, Prisoner's Dilemma and Chicken. For the moment, however, we move on to the consideration of the second fundamental idea that will be involved in the analysis of 2 × 2 ordinal games.

DEFINITION. An outcome in a 2 × 2 ordinal game is said to be a *Nash equilibrium* if neither player would gain by unilaterally changing his or her strategy.

Our formalization of 2 × 2 ordinal games makes no provision for either player actually changing his or her mind. The game is played by a single simultaneous choice of strategy (C or N), and that's the end of it. There are, however, at least two good reasons for having the concept of a Nash equilibrium at hand. First, the real world is not static; it is extremely dynamic. Hence, when we set up our models so that an outcome of a 2×2 ordinal game corresponds to a real-world event, we'll want to ask about any predictions of events to unfold suggested by the model. Second, we will later formalize this dynamic aspect of the real world, developing models that allow precisely the kind of change in choice of strategy indicated above.

An outcome that is a Nash equilibrium is one that we will think of as being stable: No one wants to upset things—at least, not *unilaterally*. We should also note that, in game theory, a Nash equilibrium is a set of strategies, not an outcome. With 2×2 ordinal games, however, there is no harm in identifying the outcome with the strategies that lead to it since they are unique.

Examples of Nash equilibria will again occur in our presentations of Prisoner's Dilemma and Chicken. Nash equilibria, by the way, are named for John Nash, whose remarkable story was told in the book and movie entitled *A Beautiful Mind*.

4.4 PRISONER'S DILEMMA AND THE ARMS RACE

Consider the following (hypothetical) situation. Two suspects are charged with having jointly committed a crime. They are then separated and each is told that both he and his alleged accomplice will be offered the choice between remaining silent (as permitted by the Miranda Decision) or confessing. Each is also told that the following penalties will then be applied:

1. If both choose to remain silent, they will each get a one-year jail term based on a sure-fire conviction on the basis of a lesser charge.

2. If both confess, they will each get a five-year prison sentence.

3. If one confesses and one remains silent, then the confessor will be regarded as having turned state's evidence and he or she will go free. The other, convicted on the testimony of the first, will get a ten-year sentence.

The question of interest then becomes the following. Assume you are one of the suspects and your sole interest is in minimizing the length of time you will spend in jail. Do you remain silent or confess?

Intuition may yield a response such as: "I wish I knew what my partner is doing." Surprisingly, this is wrong. What your partner is doing is irrelevant; you should confess. Let's see why this is true. There are two cases to consider. That is, your partner will either remain silent or confess. In the former case (remaining silent), your confession gets you off scot-free as opposed to the one-year jail term you'd get if you also remained silent. In the latter case (where he confesses), your confession gets you off with five years as opposed to the ten years you'd get for remaining silent in the face of his confession. Hence, confessing gets you a shorter jail sentence than remaining silent *regardless* of whether your partner confesses or remains silent.

The same reasoning applies to your partner. Thus, rational action (in terms of self-interest) leads to both you and your partner confessing and hence serving five years each. What is paradoxical here, however, is the observation that if both of you remained silent, you would serve only one year each and thus both be better off.

The above situation lends itself naturally to a description via a 2×2 ordinal game where "cooperation" (C) corresponds to "remaining silent" and "noncooperation" (N) to "confessing." (Think of "cooperating" as referring to cooperation with your partner as opposed to cooperation with the D.A.) Then Row, for example, ranks the outcomes from best (4) to worst (1) as:

4: NC – Row confesses and Column is silent: Row goes free.

3: CC – Row is silent and Column is silent: Row gets one year.

2: NN – Row confesses and Column confesses: Row gets five years.

1: CN – Row is silent and Column confesses: Row gets ten years.

Column's ranking is the same for CC and NN, but with "1" and "4" reversed for NC and CN. Hence the 2×2 ordinal game that models this situation is precisely the example from **Section 4.2** (duplicated in Figure 5 on the next page in our present hypothetical scenario).

Thus, in the hypothetical situation involving "prisoners" both should choose to confess even though both would benefit if both

		Column	
		C (silent)	N (confess)
Row	C (silent)	(3, 3)	(1, 4)
	N (confess)	(4, 1)	(2, 2)

FIGURE 5

remained silent. The following proposition simply formalizes this in the context of the 2×2 ordinal game Prisoner's Dilemma.

PROPOSITION. *The strategy N is a dominant strategy for both Row and Column in the game Prisoner's Dilemma.*

PROOF. We prove that N is dominant for Row; the proof for Column is analogous. Thus, we must show that, regardless of what Column does, N is a better choice for Row than is C. Column can do two things; we consider these separately.

Case 1: Column chooses C In this case, Row's choice of N yields an outcome for Row of "4" from (4, 1) as opposed to "3" from the outcome (3, 3) that would have resulted from Row's choice of the strategy C.

Case 2: Column chooses N In this case, Row's choice of N yields an outcome for Row of "2" from (2, 2) as opposed to "1" from the outcome (1, 4) that would have resulted from Row's choice of the strategy C.

Thus, we've shown that, regardless of what Column does (i.e., whether we're in case 1 or case 2), N yields a better outcome for Row than does C. This completes the proof.

The paradoxical nature of Prisoner's Dilemma is now at least partially formalized: both Row and Column have dominant strategies leading to a (2, 2) outcome that is strictly worse—for both—than the (3, 3) outcome available via mutual cooperation. Such mutual cooperation could be induced by adding additional structure to the model—threats, repeated plays of the game (see Exercises 27–29), etc.—but in the absence of such things, how does one argue against the use of a dominant strategy?

The above proposition illustrates how to prove that a given strategy is dominant for a given player. What it does not illustrate, however, is how one finds the strategies (if they exist) that are dominant. With a little experience, one can do this just by staring at the preference matrix. A better method, however, is given in the exercises at the end of the chapter.

Another paradoxical aspect of Prisoner's Dilemma is the fact that not only does the (2, 2) outcome arise from the use of dominant strategies, but, once arrived at, it is incredibly stable. This stability is formalized in the following.

PROPOSITION. *The outcome (2, 2) is a Nash equilibrium in the game Prisoner's Dilemma.*

PROOF. If Row unilaterally changes from N to C, the outcome would change from (2, 2) to (1, 4) and, in particular, be worse for Row (having gone from "2" to "1" in the first component). Similarly, if Column unilaterally changes from N to C, the outcome would change from (2, 2) to (4, 1) and be worse for Column in exactly the same way as it was for Row (having now gone from "2" to "1" in the second component). This completes the proof.

For our purposes, the importance of Prisoner's Dilemma is as a simple model of some significant political events. We consider one such example now and several more potential examples in the exercises at the end of the chapter. Our treatment is largely drawn from Brams (1985a); the reader is referred there for more background and analysis.

The U.S.–Soviet arms race of the 1960s, 1970s, and 1980s is a natural candidate for game-theoretic modeling since the actions of both countries certainly influenced and were influenced by those of the other. There is also an intractability here that, at the time, seemed to defy rationality in light of the economic burdens being imposed on both countries. Our goal here is to model the arms race as a simple 2×2 ordinal game (which turns out to be Prisoner's Dilemma) and thus explain some of the intractability as being a consequence of the structure of preferences as opposed to irrationality on the part of either country.

The model will be an enormous oversimplification of the real-world situation. It will ignore a number of admittedly important factors (such as the political influence of the military-industrial complex in each country and the economic role played by military expenditures in avoiding recessions) and focus instead on the following underlying precepts:

1. Each country has an option to continue its own military buildup (to arm), or to discontinue the buildup and begin a reduction (to disarm).

2. Both countries realize that the (primarily economic) hardships caused by an arms race make a mutual decision to arm less desirable than a mutual decision to disarm.

3. Each country would prefer military superiority to military parity. (Notice here that, although we are talking about the 1960s, 1970s, and 1980s, this may well have been false by the late 1980s.)

Given these (and the obvious least preference for military domination by the other country), we see that each country would rank the four possibilities, from most preferred to least preferred, as follows:

4. Military superiority (via the other's unilateral disarmament).

3. Mutual disarmament (parity without economic hardships).

2. An arms race (parity, but with economic hardships).

1. Military inferiority (via its own unilateral disarmament).

Thus, if we let the Soviets play the role of "Column" and the U.S. the role of "Row", with "cooperate" (C) corresponding to "disarm" and "noncooperate" (N) corresponding to "arm," the 2 × 2 ordinal game modeling this situation turns out to be (a relabeled version of) Prisoner's Dilemma (Figure 6 on the next page).

Again we see the paradox: Both countries prefer mutual disarmament—the (3, 3) outcome—to an arms race—the (2, 2) outcome. However, both countries have a dominant strategy to arm, and thus individual rationality produces the arms race no one wants.

		Soviet Union	
		Disarm	Arm
Arms Race as Prisoner's Dilemma U.S.	Disarm	(3, 3)	(1, 4)
	Arm	(4, 1)	(2, 2)

FIGURE 6

■ 4.5 CHICKEN AND THE CUBAN MISSILE CRISIS

The 2×2 ordinal game known as "Chicken" is named after the less than inspiring real-world (one would like to think hypothetical) "sport" in which opposing drivers maintain a head-on collision course until at least one of them swerves out of the way. The one who swerves first loses. Ties can occur.

In modeling Chicken as a 2×2 ordinal game, we identify the strategy "swerve" with cooperation, and "don't swerve" with noncooperation. The difference between Chicken and Prisoner's Dilemma is the interchange of preference "2" and preference "1" for both players. That is, in Prisoner's Dilemma, your least preferred outcome is a combination of cooperation on your part met by noncooperation on the part of your opponent. In Chicken, however, this outcome—although not all that great—is strictly better than mutual noncooperation. The matrix notation for Chicken is shown in Figure 7 on the next page.

Notice that the game, like Prisoner's Dilemma, is symmetric (i.e., seen the same way from the point of view of Column or Row). In terms of dominant strategies and Nash equilibria, we have the following:

PROPOSITION. *In the game of Chicken, neither Row nor Column has a dominant strategy, but both (2, 4) and (4, 2) are Nash equilibria (and there are no others).*

PROOF. We shall begin by showing that C is not a dominant strategy for Row. To do this, we must produce a scenario in which N yields a better result for Row than does C. Consider the scenario where Column

	Column	
	C (swerve)	N (don't swerve)
C (swerve)	(3, 3)	(2, 4)
N (don't swerve)	(4, 2)	(1, 1)

Row

FIGURE 7

chooses C. Then, a choice of N by Row yields (4, 2) and hence "4" for Row while a choice of C by Row yields (3, 3) and hence only "3" for Row. Thus, N is a strictly better strategy for Row than C in this case (i.e., in this scenario), and so C is not a dominant strategy for Row. Similarly, one can prove that N is not a dominant strategy for Row, and that neither C nor N is a dominant strategy for Column.

To show that (2, 4) is a Nash equilibrium, we must show that neither player can gain by unilaterally changing his or her strategy. We'll show it for Row; the proof for Column is completely analogous. If Row unilaterally changes from C to N, then the outcome would change from (2, 4) to (1, 1) and, in particular, be worse for Row (having gone from "2" to "1" in the first component). This shows that (2, 4) is a Nash equilibrium. The proof that (4, 2) is a Nash equilibrium is left to the reader, as is the proof that there are no others.

Comparing the above proposition with the one in **Section 4.4**, we see the fundamental difference between Prisoner's Dilemma and Chicken:

1. In Prisoner's Dilemma, both players have a dominant strategy, and so there is an expected (although paradoxically unfortunate) outcome of (2, 2). Moreover, because this outcome is the result of dominant strategies, it is also a Nash equilibrium (see Exercise 11), and thus (intuitively) stable.

2. In Chicken, there is no expected outcome (i.e., no dominant strategies) although (3, 3) certainly suggests itself. This outcome, however, is unstable (not a Nash equilibrium), and only

a fear of the (1, 1) outcome would prevent Row and Column from trying for the (4, 2) and (2, 4) outcomes.

Thus, instability and flirtations with noncooperation tend to characterize those real-world situations most amenable to game-theoretic models based on Chicken.

In October 1962, the United States and the Soviet Union came closer to a nuclear confrontation than perhaps at any other time in history. President John F. Kennedy, in retrospect, estimated the probability of nuclear war at this time to be between one-third and one-half. The event that precipitated this crisis was the Soviet installation of medium and intermediate range nuclear missiles in Cuba, and the subsequent detection of this by U.S. intelligence. History now refers to this event as the Cuban missile crisis.

The events that actually unfolded ran as follows. By mid-October 1962, the Central Intelligence Agency had determined that Soviet missiles had been installed in Cuba and were within ten days of being operational. Kennedy convened a high-level executive committee that spent six days in secret meetings to discuss Soviet motives, decide on appropriate U.S. responses, conjecture as to Soviet reaction to U.S. responses, and so on. The final decision of this group was to immediately put in place a naval blockade to prevent further shipments of missiles, while not ruling out the possibility of an invasion of Cuba to get rid of the missiles already there. Khrushchev, on behalf of the Soviets, responded by demanding that the United States remove its nuclear missiles from Turkey (a demand later granted—although not publicly—by Kennedy), and promise not to invade Cuba (a demand granted by Kennedy). The Soviets then withdrew all their missiles from Cuba.

Much has been written about the Cuban missile crisis and game-theoretic models thereof. Our purpose here is to present two of the simplest such models based on the game of Chicken. The first is from Brams (1985a, 1985b). The difference in the two models lies in the specification of alternatives available to the players. It may be that the former model represents more of a U.S. point of view of the situation and the latter more of a Soviet point of view. Figure 8 on the next page presents the former.

		Soviet Union	
		Withdraw missiles	Maintain missiles
Cuban Missile Crisis as Chicken **U.S.**	Blockade	(3, 3)	(2, 4)
	Airstrike	(4, 2)	(1, 1)

FIGURE 8

It should be pointed out that Brams (1985a, 1985b) embellishes the model in several ways (e.g., by consideration of deception, threats, sequential nature of the events, etc.), as well as considering a different ranking of the alternatives by the players.

The actual Soviet motives for the installation of the missiles in the first place are apparently still not known, although the fear of a U.S. invasion of Cuba may well have played a role. For more on this, see Brams (1993). If we accept this as a primary issue in the minds of the Soviets, then the game (especially as perceived by the Soviets) may have been as in Figure 9 below. Notice that the underlying 2×2 ordinal game is again Chicken. Thus, in both models, the structure of the underlying game sheds light on the tensions of these dramatic times in the early 1960s.

		Soviet Union	
		Withdraw missiles	Maintain missiles
U.S.	Give up option to invade Cuba	(3, 3)	(2, 4)
	Invade Cuba	(4, 2)	(1, 1)

FIGURE 9

■ 4.6 THE YOM KIPPUR WAR

In October 1973, the Yom Kippur War pitted Israel against a combination of Egyptian and Syrian forces. Israel quickly gained the upper hand, at which point the Soviet Union made it known that it was seriously considering intervening on behalf of Egypt and Syria. The Soviets also made it known that they hoped the United States would cooperate in what they referred to as a peace initiative. On the other hand, they were certainly aware of the U.S. option to frustrate this Soviet initiative by coming to the aid of Israel.

The above situation, again in very simplistic terms, suggests a 2 × 2 ordinal game model (Figure 10 below), where the rankings of preferences have not been filled in yet.

The question now becomes: How did the Soviet Union and the United States rank the different outcomes, and was each aware of the other's preferences? History suggests that the Soviets were convinced the preferences were as shown in Figure 11 on the next page.

Notice that this is not Prisoner's Dilemma since the United States is ranking CN ahead of NN. (That is, if the Soviets choose N, the United States would rather choose C than N.) Why would the Soviets think the United States would not respond to Soviet intervention by intervention of its own? The answer here seems to lie with the U.S. political situation at home at this time. The Watergate scandal was creating what was perceived as a "crisis of confidence" in the U.S. political

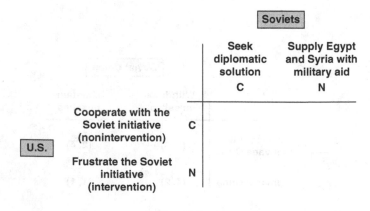

FIGURE 10

		Soviets	
		Seek diplomatic solution	Supply Egypt and Syria with military aid
		C	N
Cooperate with the Soviet initiative (nonintervention)	C	(3, 3)	(2, 4)
Frustrate the Soviet initiative (intervention)	N	(4, 1)	(1, 2)

U.S.

FIGURE 11

arena. Hence, the Soviets thought that a decision to give military aid to Egypt and Syria would not be met with an appropriate response from the United States.

President Nixon, however, realized exactly how the Soviets perceived the situation, and the consequences of this perception (see Exercise 1 at the end of the chapter). Hence, his immediate goal became that of convincing the Soviets that the correct model was, in fact, Prisoner's Dilemma as shown in Figure 12 below.

Nixon's method of accomplishing this was to place the U.S. forces on worldwide alert—one of only about half a dozen times that nuclear threats have been employed by the United States. This move (since

		Soviets	
		Seek diplomatic solution	Supply Egypt and Syria with military aid
		C	N
Cooperate with the Soviet initiative (nonintervention)	C	(3, 3)	(1, 4)
Frustrate the Soviet initiative (intervention)	N	(4, 1)	(2, 2)

U.S.

FIGURE 12

characterized by then–Secretary of State Henry Kissinger as a "deliberate overreaction") seems to have been effective in convincing the Soviets that Prisoner's Dilemma was, in fact, the correct model for U.S. and Soviet preferences in this situation.

The astute reader may well be asking the following question: Did Nixon actually gain anything by convincing the Soviets that the game was Prisoner's Dilemma? This is the issue we take up in the next section.

■ 4.7 THE THEORY OF MOVES

Recall that our basic 2×2-ordinal games are played by a single simultaneous choice of strategy (C or N) by both players. An outcome is then decided, and that's the end of it. In particular, as game-theoretic models of real-world situations, these 2×2 games are about as simple as one could hope for. The price paid for this simplicity, however, is a loss of the dynamics found in the real world.

As a particular example of the kind of loss referred to above, let's return to the considerations in **Section 4.6** and the Yom Kippur War. Recall that Nixon placed U.S. forces on worldwide alert in order to convince the Soviets that the game being played was really Prisoner's Dilemma. But now, let's face up to a fundamental difficulty with this game-theoretic model of that particular conflict:

It simply doesn't work.

In what sense does the above model fail to work? The answer: It fails to explain what actually happened. That is, the existence of dominant strategies—for intervention in this case—should have resulted in mutual noncooperation between the United States and the Soviet Union. But, in fact, neither chose to intervene and so we wound up at the (3, 3) outcome (which is also unstable in the sense of not being a Nash equilibrium). What is wrong with our model, and can it be modified to more faithfully reflect reality?

The most obvious place to look for a shortcoming in the Prisoner's Dilemma model of intervention in the Yom Kippur War is with the preference rankings we assigned. But this is probably not where the problem lies in this particular case. The problem is even more basic than the choice of preferences.

There is a very fundamental way in which a 2 × 2-ordinal game differs from the situation in which the United States and the Soviet Union found themselves in 1973. This difference rests in what we might call the "starting position." In a 2 × 2-ordinal game, the starting position is completely neutral—neither C nor N has any predetermined favored status. But in the real-world situation of the Yom Kippur War, the starting position was clearly one of mutual nonintervention. Hence, the United States and the Soviet Union were already at the (3, 3) outcome and the question was whether or not either side should change its status quo strategy of C (nonintervention) to N (intervention).

From this point of view, the game certainly does start to explain the events that unfolded. That is, the (3, 3) outcome is definitely not stable (i.e., not a Nash equilibrium) and so it is certainly rational for each side to try to find out the resolve of the other with respect to responding to a switch from C to N. This, of course, is exactly what the Soviets did, and Nixon's response was designed to convey a very exact message regarding this resolve. Thus, a better way to use the 2 × 2-ordinal preference matrix in modeling this particular situation is to consider a new kind of game where a starting position is determined in some way, and then each player has the option of changing strategy. This leads us directly to (a slightly modified version of) the so-called theory of moves introduced by Brams and Wittman (1981) and extensively pursued in Brams (1994). The precise definitions and rules are as follows.

To each 2 × 2-ordinal preference matrix (like that for Chicken or Prisoner's Dilemma) we associate two "sequential games"—one in which Row goes first and one in which Column goes first. We'll describe the former version; the latter is completely analogous. Suppose we have a fixed 2 × 2-ordinal preference matrix. The "sequential game" (with Row going first) proceeds as follows.

Step 1: Both players make an initial simultaneous choice of either C or N. This determines what we will call an *initial position* of the game.

Step 2: Row has the choice of leaving things as they are ("staying"), or changing his strategy.

Step 3: Column has the same choices as did Row in step 2.

They continue alternately. The game ends if any one of the following situations occurs:

1. It is Column's turn to move and the position of the game is (-, 4).

2. It is Row's (second or later) turn and the position of the game is (4, -). Thus, if the initial position is (4, -), the game does not immediately end.

3. Either Row or Column chooses to stay, with the one exception to this being that an initial "stay" by Row does not end the game; we give Column a chance to move even if Row declines the chance to switch strategy on his first move.

Notice that the effect of rules 1 and 2 is to build a bit of rationality into the rules of the game. This guarantees that certain games will terminate and thus be susceptible to the kind of tree analysis we want to do.

The outcome at which a game ends is called the *final outcome* and this (alone) determines the payoffs.

The analogue of a Nash equilibrium in the present context is given by the following.

DEFINITION. An outcome is called a *non-myopic equilibrium when Row goes first* if sequential rational play in the game described above results in that outcome being the final outcome any time it is chosen as the initial position. The notion of a "non-myopic equilibrium when Column goes first" is defined similarly.

DEFINITION. A *non-myopic equilibrium* is an outcome that is both a non-myopic equilibrium when Row goes first and a non-myopic equilibrium when Column goes first.

We will analyze rational play in the kind of sequential game described above by what is called "backward induction" or, more informally, "pruning the tree." This so-called "game-tree analysis" begins at any point in the tree where the next move will definitely end the game (according to the rules). Assuming the player about to make this final move is rational—meaning that he will choose, of the two possible final outcomes resulting from his move, the one that is better for him—we can eliminate from consideration (and from the tree) the potential move that will be rejected by this player. The result of our eliminating this move is a smaller tree that nevertheless represents the same game (assuming, as we are, that players are rational). Continuing this "pruning" eventually reveals the optimal sequence of moves that would be chosen by rational players.

We will illustrate backward induction in the sequential version of Prisoner's Dilemma where Row goes first. It will turn out, in fact, that both the (2, 2) and (3, 3) outcomes are non-myopic equilibria. The (3, 3) outcome, however, is the result of a dominant strategy in the theory of moves version of Prisoner's Dilemma.

Our method of analyzing the theory of moves version of Prisoner's Dilemma will be to consider separately the four possible initial positions in the game. For each, we'll do a game-tree analysis and find the corresponding final outcome. This will immediately show that (2, 2) and (3, 3) are the only non-myopic equilibria. Further analysis at this point will then yield the additional claim about the dominant strategy. Recall that Row is going first.

Case 1: The Initial Position is (3, 3) in Prisoner's Dilemma

The tree of possibilities, displayed in Figure 13 on the next page, is constructed in the following way from the 2 × 2 ordinal game (which is also reproduced in the small box within Figure 13).

1. The top node is the initial position, which is (3, 3) in this case.

2. Row gets to move first and has a choice between staying at (3, 3) or switching strategies from "C" to "N" and thus moving the position of the game to (4, 1). This explains the two nodes labeled "stay at (3, 3)" and "(4, 1)" on the level of the tree just below the top (3, 3) node. Notice that to the far left of these two nodes is the word *Row*, indicating that the choice between these two is being made by Row.

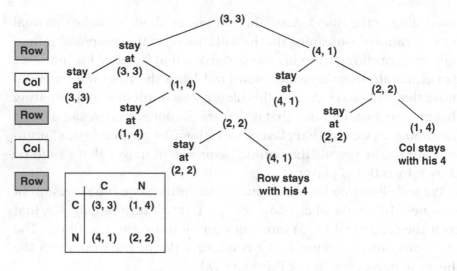

FIGURE 13

3. Column gets to move next. Recall that even if Row chooses to stay on the initial move, the game does not end. Thus, if Row chooses to stay at (3, 3), Column could also stay, ending the game at a final outcome of (3, 3), or Column could switch his strategy from "C" to "N" and thus move the position of the game from (3, 3) to (1, 4). Similarly, if Row had moved the game to (4, 1) on his first move, Column would have a choice between staying there, and ending the game at a final outcome of (4, 1), or switching strategies from "C" to "N" and thus moving the position of the game from (4, 1) to (2, 2).

4. Column and Row thus continue to alternate moves. Notice that Row is controlling the "vertical movement among outcomes" and Column is controlling "horizontal movement among outcomes."

5. Notice also that the game is finite, since the position of the game becomes (1, 4) at a time when it is Column's turn to move (thus guaranteeing a stay at his "4" by Column according to the rules) and the position of the game becomes (4, 1) at a time when it is Row's turn to move. (Not all games like this are finite—see Exercise 5.)

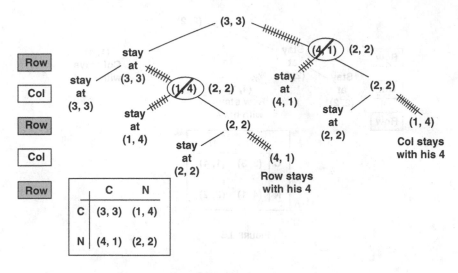

FIGURE 14

For the game-tree analysis of rational play we start at the bottom-most nodes and work our way up the tree, transferring outcomes labels up and "X-ing out" the position of the game that will not be passed through on the way to the final outcome. This is illustrated in Figure 14 above. Note, for example, that starting at the lower left part of the tree, Column has a choice between staying at (2, 2) or moving to (4, 1) where Row will definitely stay. Since Column prefers the "2" from "(2, 2)" to the "1" from "(4, 1)", the option to move will be rejected as is indicated by the "railroad tracks." Moving one level higher on that same side of the tree, we see that Row has a choice between staying at (1, 4) and getting his worst outcome, or moving to (2, 2) which will turn out to be the final outcome. Clearly he does the latter and so we "X-out" the edge leading to "stay at (1, 4)" and we replace the temporary (1, 4) label by the (2, 2) that we now know will be the final outcome if the game reaches this position.

Conclusion The game-tree analysis from Figure 14 shows that rational play dictates an initial choice to stay at (3, 3) by Row, followed by Column's choice to also stay and to thus let (3, 3) be the final outcome as well as the initial position. Hence, (3, 3) is a non-myopic equilibrium when Row goes first.

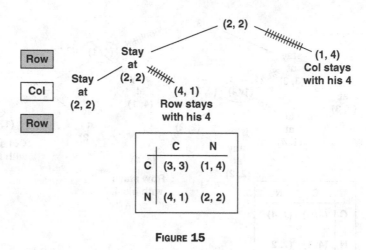

FIGURE 15

For the three remaining cases, we will present only the analogues of Figure 14 and the conclusions they yield.

Case 2: The Initial Position is (2, 2) in Prisoner's Dilemma

Conclusion The outcome (2, 2) is a non-myopic equilibrium when Row goes first (Figure 15 above). In fact, with (2, 2) as the initial position, rational play dictates that Row will choose to stay as will Column.

Case 3: The Initial Position is (1, 4) in Prisoner's Dilemma

Conclusion If the initial position is (1, 4) in Prisoner's Dilemma, then Row will switch strategies, thus moving the outcome to (2, 2) (Figure 16 on the next page). Column will then choose to stay and the game will end at (2, 2). Intuitively, this says that if Column is being aggressive and Row is not, then Row will respond to this by also being aggressive and that's where things will stay.

Case 4: The Initial Position is (4, 1) in Prisoner's Dilemma

Conclusion If the initial position is (4, 1) in Prisoner's Dilemma, then Row will switch strategies and thus move the outcome to (3, 3) (Figure 17, next page). Column will then choose to stay. Intuitively,

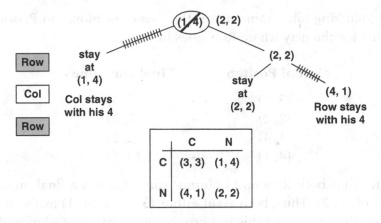

FIGURE 16

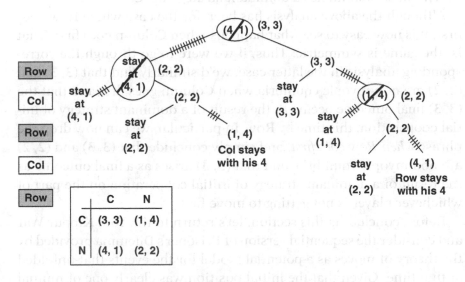

FIGURE 17

if Row is being aggressive and Column is not, then Row realizes that if he does not back off to a nonaggressive stance, then Column will become aggressive and the (2, 2) stalemate will prevail instead of the (3, 3) compromise.

The following table summarizes the theory of moves in Prisoner's Dilemma for the play where Row goes first.

Initial Positon		Final Outcome
(3, 3)	$\longrightarrow$	(3, 3)
(2, 2)	$\longrightarrow$	(2, 2)
(1, 4)	$\longrightarrow$	(2, 2)
(4, 1)	$\longrightarrow$	(3, 3)

Notice that both Row and Column want (3, 3) as a final outcome instead of (2, 2). Thus, both want either (3, 3) or (4, 1) as the initial position. However—and this is a crucial observation—Column alone can guarantee this simply by choosing C as his initial strategy. Then, if Row chooses C we start at (3, 3) and if Row chooses N, we start at (4, 1). Thus, Column has a dominant strategy of "C."

Although the above analysis has been for the case where Row goes first, it is now easy to see what happens when Column goes first. That is, the game is symmetric. Thus, if we were to go through the corresponding analysis in the latter case, we'd similarly find that (3, 3) and (2, 2) are non-myopic equilibria when Column goes first and that the (3, 3) final outcome occurs as the result of a dominant strategy of initial cooperation, this time by Row. In particular, we can now drop the phrase *when Row goes first,* and simply conclude that (3, 3) and (2, 2) are non-myopic equilibria, and that (3, 3) arises as a final outcome as the result of a dominant strategy of initial cooperation on the part of whichever player is not getting to move first.

Before concluding this section, let's return to the Yom Kippur War and consider the sequential version of Prisoner's Dilemma provided by the theory of moves as a potential model for the events that unfolded at that time. Given that the initial position was clearly one of mutual nonintervention—the (3, 3) outcome in our model—then the model accurately predicts exactly what happened. That is, neither side elected to change its initial choice of strategy. Notice that since (3, 3) is a non-myopic equilibrium, the question of whether the United States or the Soviet Union is "designated" as going first doesn't arise. However, it seems clear that in the analysis of the situation by both sides, the Soviets were more likely to play this role.

■ 4.8 CONCLUSIONS

In this chapter, we've introduced 2×2 ordinal games in general as well as the two most interesting examples of such. The first—Prisoner's Dilemma—is one in which both players (independently) have dominant strategies leading to a (2, 2) outcome that both consider inferior to the (3, 3) outcome that is available. The (2, 2) outcome also turns out to be stable in the sense of being a Nash equilibrium (where neither player can gain by unilaterally changing his or her strategy). We also presented in this chapter the classic application of Prisoner's Dilemma as a model of the U.S.-Soviet arms race of the 1960s, 1970s, and 1980s.

The second 2×2 ordinal game introduced in this chapter is Chicken. This game is quite different from Prisoner's Dilemma in the sense that Prisoner's Dilemma has an expected, although paradoxically unfortunate, (2, 2) outcome, while there are no dominant strategies in Chicken, although (2, 4) and (4, 2) are stable outcomes (arrived at only by flirting with the disastrous (1, 1) outcome). As an application of Chicken, we constructed two different models of the Cuban missile crisis. The difference between these models is in the choice of strategies available to the two players.

We also considered the Yom Kippur War, and observed that the naïve 2×2 ordinal game-theoretic model simply did not work in the sense of predicting what actually took place. With this in mind, we turned, in **Section 4.7**, to a more complicated game involving the so-called theory of moves. In particular, the theory of moves explains why an initial position of mutual cooperation on a Prisoner's Dilemma game board will persist, even when both sides have the opportunity to (alternately) change strategies.

EXERCISES

1. Suppose Row ranks the four possible outcomes, from best to worst, in a 2×2 ordinal game as CN, CC, NC, NN and Column ranks the four, again from best to worst, as CC, NN, NC, CN.

    (a) Set up the 2×2 matrix (as in Figure 2 in **Section 4.2**) giving Row's preference ranking.

   (b) Set up the 2 × 2 matrix (as in Figure 3 in **Section 4.2**) giving Column's preference ranking.

   (c) Express all this information in a single 2 × 2 matrix (as in Figure 4 in **Section 4.2**).

2. Write out the proof that N is a dominant strategy for Column in Prisoner's Dilemma.

3. Show that C is a dominant strategy for (a) Row and (b) Column in the following game.

		Column	
		C	N
Row	C	(3, 4)	(4, 2)
	N	(1, 3)	(2, 1)

4. In the following 2 × 2 ordinal game:
   (a) Show that C is *not* a dominant strategy for Row.
   (b) Show that N is *not* a dominant strategy for Row.
   (c) Show that C is *not* a dominant strategy for Column.
   (d) Show that N is *not* a dominant strategy for Column.

		Column	
		C	N
Row	C	(2, 3)	(3, 1)
	N	(4, 2)	(1, 4)

5. In this chapter and in the exercises so far, we have dealt with the issue of how to prove that a given strategy is dominant in a particular 2 × 2 ordinal game. We have not yet addressed the question of how one *finds* a dominant strategy if one is handed a 2 × 2 ordinal game.

We illustrate one such procedure here. Consider, for example the following game:

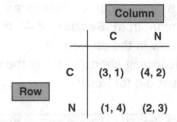

We shall first check to see if Row has a dominant strategy. Our starting point (regardless of the game) is the following chart (where we have filled in only the part of the chart that does not require looking at the game in question):

Column's Choice	Row's Best Response
C	
N	

For this particular game, we can see that if Column chooses C, then the outcome will be either (3, 1) or (1, 4), and Row would certainly prefer the "3" from (3, 1) to the "1" from (1, 4). Thus, Row's best response to a choice of C by Column is C, since this is what yields the outcome (3, 1). A similar analysis when Column chooses N shows that Row's best response is also C in this case. Thus, the rest of the chart can be filled out as follows:

Column's Choice	Row's Best Response
C	C (because 3 > 1)
N	C (because 4 > 2)

From this we can conclude that Row has a dominant strategy of C. Notice, however, that the above chart is a poor excuse for a proof that C is a dominant strategy for Row. That is, a proof is a convincing argument, and the above chart conveys little to anyone who does not already understand the material. On the other hand, the chart (together with the preference matrix) should make it easy for the reader to:

(a) Write down a proof (with sentences as in the proof for Prisoner's Dilemma from **Section 4.4**) that C is a dominant strategy for Row in the above game.

(b) Fill out the following chart (which is the analogue for Column of what we just did for Row):

Row's Choice	Column's Best Response
C	
N	

(c) Use what you found from the chart in part (b) to prove that Column has no dominant strategy. (This should look like the proof for Chicken in **Section 4.5**.)

Notice that in filling out these charts, there are four possibilities for what can occur below the "Best Response" label:

$$C \quad C \quad N \quad N$$
$$C \quad N \quad C \quad N$$

In the first case, C is a dominant strategy, and in the last case, N is a dominant strategy. In the second case, the optimal strategy suggested is called "tit-for-tat." In the third case, it is called "tat-for-tit."

6. Find the dominant strategies in the following game and prove that they are, in fact, dominant.

		Column	
		C	N
	C	(2, 1)	(1, 2)
Row			
	N	(4, 3)	(3, 4)

7. Determine if there are any dominant strategies in the following game.

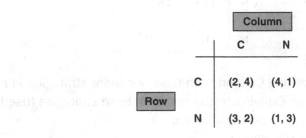

		Column	
		C	N
Row	C	(2, 4)	(4, 1)
	N	(3, 2)	(1, 3)

8. Extend what is done in Exercise 5 to answer the following: Does Column have a dominant strategy in the following 2×3 game where Column has three choices: C, N, and V? (Intuition: think of V as very uncooperative.) Each ranks the six possible outcomes from 6 (best) to 1 (worst).

		Column		
		C	N	V
Row	C	(5, 4)	(3, 5)	(2, 6)
	N	(6, 1)	(4, 2)	(1, 3)

9. Suppose that CC is a (4, 4) outcome in a 2×2 ordinal game. Does this guarantee that C is a dominant strategy for both Row and Column? (Either explain why it does, or find a 2×2 ordinal game showing that it need not.)

10. In the following game:
    (a) Show that (2, 3) is a Nash equilibrium.
    (b) Show that (4, 2) is not a Nash equilibrium.
    (c) Is (3, 4) a Nash equilibrium? (Why or why not?)
    (d) Is (1, 1) a Nash equilibrium? (Why or why not?)

		Column	
		C	N
Row	C	(2, 3)	(4, 2)
	N	(1, 1)	(3, 4)

11. Suppose that Row and Column both have dominant strategies in a 2×2 ordinal game. Explain why the result of these strategies (used simultaneously) is a Nash equilibrium.

12. Consider the following game:

	C	N
C	(2,2)	(3,3)
N	(1,4)	(4,1)

   (a) Prove that Row has no dominant strategy.
   (b) Prove that Column has no dominant strategy.
   (c) Prove that this game has no Nash equilibrium.

13. Consider a two-player game in which the players simultaneously show a penny, either heads up or tails up. If both players show heads, then both players lose their pennies to a lucky third party, and if both players show tails, each player keeps his or her own penny. If both players show different sides, then the player who shows heads gets both coins.

   (a) Write down the two-by-two matrix for this game.
   (b) Is Chicken or Prisoner's Dilemma or neither a model for this game?
   (c) Do the players have a dominant strategy?
   (d) Is there a Nash equilibrium?

14. Consider the following game:

	C	N
C	(2,3)	(4,2)
N	(1,1)	(3,4)

   (a) Prove that Row has a dominant strategy. What is it?
   (b) Prove that Column has no dominant strategy.
   (c) Are there any Nash equilibria?

15. In long distance cycling races, drafting is a frequent phenomenon. When one cyclist rides behind someone else, the wind resistance is cut, and it is much easier to pedal; experts suggest that the cyclist in back can save between 20% and 40% of his energy during the race. Top cycling teams often use this strategy; the team players take turns riding in front of the team leader who then has a better chance of winning the race.

   Suppose that two friends enter a cycling race, and at one point near the end of the race, the two cyclists find themselves a good distance ahead of the rest of the group. Their energy is lagging, and if both riders continue to work alone, the rest of the pack will soon catch up, and neither will win. If the two take turns drafting, then they will remain ahead of the pack for awhile; it's possible that one of the two will win, but it's more likely that they will both tire enough that someone else passes them in the end. If either cyclist pulls just ahead of his friend, however, allowing him to draft the rest of the race, then the two will remain ahead of the pack, and the cyclist in back will certainly have the energy to pull ahead in the last leg and win. Each cyclist would prefer to win the race, but would rather see his friend win than a stranger. Model this scenario with a 2×2 ordinal game, and determine what, if anything, the model predicts will happen. Is the game Prisoner's Dilemma, Chicken, or neither?

16. Kathryn and Nadia each plan to throw a New Year's Eve party; each one has a back-up date as well, and the two back-up dates do not conflict. Ideally, Kathryn hopes that she can throw the New Year's Eve party, and that Nadia will choose a different date. But if that doesn't happen, she really wants to be able to attend Nadia's party, even though she'll be very jealous if Nadia's party is on New Year's and she has to choose a different date.

(a) If Nadia feels the same way as Kathryn, write down a 2×2 ordinal game that models the situation. What, if anything, does the model predict will happen?

(b) Suppose that Nadia's first priority is that her party is on New Year's Eve, and would absolutely hate it if Kathryn gets to throw the New Year's Eve party and she is forced to choose a different date. Write down a new 2×2 ordinal game that models the situation. What, if anything, does the model predict will happen?

(c) For both scenarios above, is the game Prisoner's Dilemma, Chicken, or neither?

17. Consider the following hypothetical situation. NASA plans to launch a manned vehicle into space, but the engineers feel that it is unsafe. NASA has the options to launch or not, and the engineers have the option to go public with their reservations or not. Assume that NASA's first priority is that the engineers remain silent (because NASA honestly feels that they are wrong), and, as a second priority, NASA would rather launch than not launch. Additionally, assume that the engineers have a first priority of preventing the launch, and a second priority of going public with their reservations. Model this as a 2×2 game, and, in a few sentences, explain what outcome is predicted by the existence of dominant strategies.

18. Suppose there are two colleges, both competing for the same group of students (all of whom will go to one of the two colleges). Suppose that each college knows that if one offers merit scholarships and the other doesn't, then the one that does will enroll more of the better students and more than justify the expense. However, if both offer merit scholarships, it will be costly and have no effect on which students enroll where. Model this as a 2×2 game, and, in a few sentences, explain what outcome is predicted by the existence of dominant strategies.

19. Do there exist 2×2 ordinal games with a Nash equilibrium that is not the result of dominant strategies by Row and Column? Give an example or prove that one does not exist.

20. (This requires extending what was in the text.) Find all Nash equilibria in the following 3×3 game:

		Column		
		C	N	V
	C	(1, 9)	(4, 2)	(7, 7)
Row	N	(3, 4)	(9, 3)	(5, 1)
	V	(6, 5)	(2, 6)	(8, 8)

21. Find all Nash equilibria for the following 3×3 game, and for each outcome that is not a Nash equilibrium, explain why it is not.

	C	N	V
C	(1,4)	(2,5)	(3,3)
N	(4,8)	(5,9)	(6,2)
V	(7,6)	(8,7)	(9,1)

22. Consider the Democratic primaries prior to the 2008 presidential election. Assume that Hillary Clinton and Barack Obama had a choice of waging an aggressive (negative) campaign directed at the other's weaknesses, or waging a positive campaign based on their own strengths. Assume also that each felt that negative campaigning, unless answered in kind, would be advantageous to the one doing the negative campaigning, at least as far as the primaries are concerned. Notice, however, that mutual negative campaigning will certainly put the Democratic party in a worse position for the general election than mutual positive campaigning.

    (a) Assuming that each candidate is more concerned with his or her own political success than doing what is best for the party, model this as a 2×2 game and discuss what this suggests as far as rational behavior on the part of the candidates.

(b) How does your model change if we assume that each candidate has the party's best interests in mind?

23. In Puccini's opera *Tosca,* the main characters are the beautiful Tosca, her lover Cavaradossi, and Scarpia, the chief of police. Scarpia has condemned Cavaradossi to death, but offers to spare his life (by arranging to have blanks in the guns of the firing squad) in exchange for Tosca's favors. Tosca agrees and a meeting between her and Scarpia is set (which—exercising mathematical license—we shall assume is for the same time as the execution). Tosca thus has a choice between submitting as agreed or double-crossing Scarpia (perhaps by not showing up; perhaps in some other way). Scarpia has a choice between arranging for the blanks as agreed or double-crossing Tosca by not doing so. Tosca considers having her lover spared to be more important than the issue of whether she submits or not, even though—other things being equal—she would rather not submit. Scarpia considers having Tosca submit to be more important than the issue of whether Cavaradossi is executed or not, even though—other things being equal—Scarpia would rather have him killed.

(a) Model this as a 2 × 2 ordinal game and then determine what, if anything, the model predicts will happen.

(b) Find out what happened in the opera and see if your predictions are correct.

24. The following report appeared in *The Daily Gazette* (Schenectady, NY, Sept. 25, 1993):

> OPEC's high oil output and falling prices have cost member countries about $6 billion since the spring and some countries continue to exceed production limits, the cartel said.
>
> One day ahead of a crucial meeting on Saturday, the Organization of Petroleum Exporting Countries and its dozen members were pumping about a million barrels above the ceiling of 23.6 million barrels.

To better understand this, let's consider a hypothetical version of OPEC consisting of six countries. Assume that as the number of barrels of oil produced by OPEC per day increases, the price decreases according to the following table (which is also hypothetical):

Barrels per day produced (in millions)	24	25	26	27	28	29	30
Resulting price per barrel (in dollars)	24	23	22	21	20	19	18

Suppose OPEC agrees that each of the six countries will produce four million barrels per day, even though each country has the ability to produce five million barrels per day at no additional cost to itself. Suppose also that if anyone violates the agreement, no one will know who did (but everyone will know how many countries did because of the resulting price per barrel).

Assume you are the leader of one of the six OPEC countries and you are only concerned with financial gain for your country. You have to decide whether to produce four million barrels per day or five million barrels per day.

(a) The number of OPEC countries, other than yours, who produce five million barrels per day instead of four million could be 0, 1, 2, 3, 4, or 5. For each of these six cases, determine if your country is better off financially producing five million barrels per day or four million barrels per day.

(b) Still assuming your only concern is immediate financial gain for your country, what does (a) indicate you should do and how compelling is this indication?

(c) If all six countries care only about their own immediate financial gain, what does (b) suggest will happen?

(d) Given what you said in (c), how does your country fare financially compared to how it would do if everyone (including you) stuck to the original agreement?

(e) In a well-written paragraph or two, discuss how this hypothetical scenario is similar in spirit to something that arose in our study of 2 × 2 ordinal games.

25. In 1960 William Newcomb, a physicist, posed the following problem: Suppose there are two boxes labeled A and B. You have a choice between taking box B alone or taking both A and B. God has definitely placed $1,000 in box A. In box B, He placed either $1,000,000 or nothing, depending upon whether He knew you'd

take box B alone (in which case He placed $1,000,000 in box B) or take both (in which case He placed nothing in box B). The question is: Do you take box B alone or do you take both? You can answer this if you want to, but that's not the point of this exercise. In fact, hundreds of philosophical papers have been written on this problem. Most people think the answer is obvious, although they tend to split quite evenly on which answer is obvious and which answer is clearly wrong.

(a) Give an argument that suggests you should take both boxes.

(b) Give an argument that suggests you should take box B alone.

(c) Indicate which argument you find most compelling and why.

(d) Consider the following 2 × 2 ordinal game:

		Column	
		C	N
	C	(3, 4)	(1, 3)
Row			
	N	(4, 1)	(2, 2)

Prove that Row has a dominant strategy of N. Now suppose that we change the rules of the game so that Row chooses first, and then Column—knowing what Row did—chooses second. Explain why, even though Row has a dominant strategy of N in the game with the usual rules, Row should choose C in this version of the game where Row moves first. Here is our resolution of Newcomb's problem. (There are hundreds of "resolutions" in the literature; the reader should take the authors' with the grain of salt it probably deserves.) Consider the following 2 × 2 ordinal game between God and us. God has two choices: to put $1,000 in box A and $1,000,000 in box B, or to put $1,000 in box A and $0 in box B. We also have two choices: take both boxes or take box B alone. Our ranking of the outcomes is clear, since the dollar amounts we receive for the four possible outcomes are $1,001,000; $1,000,000; $1,000; and $0. God, on the other hand, apparently regards the upper left outcome as better than the upper right outcome (rewarding us for not being greedy). Similarly, He would seem to regard the lower right outcome as better than the lower left outcome (punishing us for our greed).

		God	
		$1,000 in A $1,000,000 in B	$1,000 in A $0 in B
Us	Choose box B alone	$(3, a^+)$	$(1, a)$
	Choose both boxes	$(4, b)$	$(2, b^+)$

Notice that this game, assuming only that a^+ is greater than a and b^+ is greater than b, has the same property as the game in part (d): We have a dominant strategy of "choose both" in the usual play of the game, but, in the game where we must move first, we are better off not using this strategy.

This is the paradoxical nature of God's action being based on His knowledge of what we will do: Which game is being played—the one where we go first (and if He knows what we will do, surely this is equivalent to our already having done it), or the one where we move independently (as in the usual play of a 2×2 ordinal game)?

26. A two-player game is said to be a *somewhat finite game* if every play of the game ends after finitely many moves. "Hypergame" was created by William Zwicker in the late 1970s. It is played by two players as follows: The first move consists of Player 1 naming a somewhat finite game of his or her choice. The second move in this play of hypergame consists of Player 2 making a legitimate first move in the somewhat finite game named in move 1. Player 1 now makes a second move in the game named, and they continue to alternate until this play of the game named is completed. (In some ways, hypergame is like dealer's choice poker.)

    (a) Write down a compelling argument that hypergame *is* a some-what finite game.

    (b) Write down a compelling argument that hypergame *is not* a somewhat finite game.

    More on hypergame is readily available in Zwicker (1987).

27. Iterated Prisoner's Dilemma is a two-player game in which two players play the Prisoner's Dilemma game a fixed finite number N of times.

    (a) Determine each player's strategy when $N = 2$.

(b)   Determine each player's strategy when $N = 3$.

(c)   Explain why each player's strategy remains the same no matter how large N is.

28.   Robert Axelrod, a political scientist, organized a tournament in which participants played an iterated version of Prisoner's Dilemma, that is, the game is played a certain number of times, and the players may base their strategies in one round on their opponent's behavior in the previous round. The player who wins the most rounds is the winner. Some possible strategies are as follows.

Pure Cooperation. The player cooperates during every iteration of the game.

Pure Non-Cooperation. The player does not cooperate during every iteration of the game.

Random. The player flips a coin for every iteration of the game: if heads comes up, he cooperates, and if tails comes up, he does not cooperate.

Alternation. The player cooperates during the first round and in every other odd-numbered round, and does not cooperate in all even-numbered rounds.

Tit-for-Tat. The player cooperates during the first round of play. During all other rounds, the player uses the strategy that his opponent used during the previous round.

(a)   Suppose that two players play a 5-round Iterated Prisoner's Dilemma, and both use the Tit-for-Tat strategy. Describe the outcome of the game, that is, who wins during each of the five stages.

(b)   Suppose that two players play a 5-round Iterated Prisoner's Dilemma; Player 1 uses the Pure Non-Cooperation strategy, and Player 2 uses the Tit-for-Tat strategy. Discuss the outcome of the game.

(c)   Suppose that two players play a 5-round Iterated Prisoner's Dilemma; Player 1 uses the Alternation strategy, and Player 2 uses the Tit-for-Tat strategy. Discuss the outcome of the game.

(d)   Suppose that you are playing a 5-round Iterated Prisoner's Dilemma, and you know your opponent will use Tit-for-Tat strategy. What should you do at each stage of the game?

29.   How might a player's strategy for Iterated Prisoner's Dilemma differ if infinitely many rounds are played?

30. The ultimatum game is a two-player game, played as follows: Player 1 proposes a possible division of $1 between the two players (for example, they might split the $1 evenly between them). Only divisions requiring quarters (no dimes, nickels, or pennies) are allowed. Player 2 has two options: she can either accept the division and the dollar is split as proposed, or she can reject the division in which case neither player receives anything.

    (a) Assuming each player just wants to maximize his profit, what is Player's 2 dominant strategy? What about Player 1?

    (b) In practice, a large percentage of the people in Player 1's role offer a near 50-50 split. Compare this to your results in part (a). How might you explain this difference?

31. Suppose the Soviets think that the correct model of the Yom Kippur War is the one in Figure 11 in **Section 4.6**. Based on this model, what would the Soviets expect to happen?

32. In a few sentences each, explain the steps in the analysis pictured in Figures 15, 16, and 17 in **Section 4.7**.

33. Show that (3, 3) is a non-myopic equilibrium in the theory of moves version of Chicken.

34. Do an analysis of the theory of moves version of Chicken that is analogous to what was done for Prisoner's Dilemma.

35. Show that the theory of moves version of the following game is not finite. Assume that (2, 3) is the initial position and Row goes first.

		Column	
		C	N
Row	C	(2, 3)	(3, 1)
	N	(4, 2)	(1, 4)

Fairness

■ 5.1 INTRODUCTION

The central difficulty in solving most disputes is finding a solution that all parties involved consider "fair." Of course, fairness is a subjective issue, and is very difficult to define or quantify. Surprisingly, however, a mathematical perspective can help identify what it means for a solution to be fair and offer a variety methods or procedures for achieving a solution in many types of disputes.

We will focus primarily on fairness in two distinct realms: apportionment and fair division. According to the United States Constitution, the number of congressional representatives per state should be assigned according to the state's population. A naïve allocation of seats to states—based on the fraction of the U.S. population residing in that state—leads to an allocation in which the number of congressional representatives for a state is not a whole number; such an allocation is impossible to implement. And yet we can't just round to the nearest whole number because the sum of the seats allocated has to be a certain fixed number (435). There are, in point of fact, a number of procedures for handling this "rounding-off problem" that have been proposed and used over the years, and we consider several in **Sections 5.2** and **5.3**.

Just as the search for a perfect social choice procedure proved doomed in Chapter 1, so too does the search for a perfect method of apportionment. We illustrate this in **Section 5.4** with a weakened-but-still-striking version of the Balinski-Young impossibility theorem.

Another common type of dispute involves fairly allocating a number of goods among several parties, such as that which occurs in the distribution of marital assets in a divorce or the division of an estate among two or more heirs. In fact, similar methods can be used for disputes in which it is not a physical set of goods that is under contention but a set of issues that need to be resolved. For example, two political parties deciding on rules for a debate between candidates might argue over the issues of the length of the debate, the source of the questions, and the time allowed for initial answers and rebuttals. We will look at what it means for solutions to these types of disputes to be fair, and how to achieve such a solution.

In **Section 5.5**, we discuss fairness in the context of dispute resolution, and some criteria by which we can judge the "fairness" of a particular method of dispute resolution or fair division. In **Section 5.6**, we look at a specific method, the adjusted winner procedure, for disputes involving two parties. In **Section 5.7** we apply the adjusted winner procedure to the Israeli - Palestinian conflict in the Middle East.

··

■ 5.2 THE PROBLEM OF APPORTIONMENT

The U.S. House of Representatives has, at any given time, a fixed size—presently 435. Article 1, Section 2 of the Constitution specifies that these seats should be apportioned among the states "according to their respective numbers." This suggests that a state with 10% of the U.S. population should have 10% of the 435 seats in the House. Alas, 10% of 435 is 43.5, and a fraction of a seat is quite impossible.

A number such as 43.5, arrived at as we did in the previous paragraph, is a state's "ideal allotment" or "quota." It is the number of seats that a state would ideally have, if fractional seats were possible. A state's quota is thus calculated by multiplying the size of the House (435) by the fraction that corresponds to the percentage of the U.S. population residing in that state.

The "apportionment problem'" refers to the search for a method to replace these quotas by whole numbers in a way that is as fair and equitable as possible. Unfortunately, the naïve solution of just rounding each fraction to the nearest whole number fails because the resulting total will typically be either less than the fixed House size (leaving seats unfilled) or greater than the fixed House size (thus apportioning non-existent seats).

Alexander Hamilton, Secretary of the Treasury, proposed the first solution to the apportionment problem following the initial U.S. census in 1792. His proposal is easy to describe:

Hamilton's Method of Apportionment: Begin by rounding all quotas down to the nearest whole number and allocate seats accordingly, leaving (typically) a number of seats not yet allocated. Now hand these additional seats out, one at a time, according to the size of the fractional part of the quota (so that a state with a quota of 13.92 would get an extra seat before a state with a quota of 31.67, because .92 is greater than .67).

Hamilton's method was not used in 1792 because President George Washington vetoed the bill (the first bill in U.S. history to suffer this fate). It was, however, resurrected in 1850 and used for the next 40 years. In **Section 5.3**, we say more of the history of apportionment in the U.S., and we'll see that the choice of method to be used was often based more on political considerations than objective issues of fairness.

For the moment, let's ask what it might mean to say that a specific method, such as Hamilton's, for apportioning seats among the 50 states is "fair and equitable." Without some attempt to formalize this via desirable properties, we're back at the constitutional directive to do it "according to their respective numbers."

As a starting point, Hamilton's method possesses two properties that certainly seem, at first blush, to be obvious desiderata.

The Monotonicity Property

A method of apportionment satisfies the *monotonicity property* (or is said to be *monotone* or *monotonic*) if no state receives fewer seats than a state with less (or the same) population. That is, if state A has fewer seats than state B, then state A should have less population than state B.

The Quota Property

A method of apportionment satisfies the *quota property* (or, more briefly, satisfies *quota*) if the number of seats allotted to a state never differs from its (ideal) quota by more than one. Thus, if a state's quota is 13.92, it should receive either 13 seats or 14 seats.

Adding to the challenge of achieving fairness in apportioning seats is the fact that a census is conducted every 10 years, and so seats will typically have to be reallocated. Achieving fairness in view of such transitions turns out to be surprisingly difficult, and Hamilton's method comes up a bit short. In particular, it fails to satisfy the following (as we will later demonstrate).

The Population Property

A method of apportionment satisfies the *population property* (or *avoids the population paradox*) if, following a census, no state should gain population and lose a seat while some other state loses population and gains a seat.

It turns out that there is no shortage of apportionment methods that satisfy the population property. These are the so-called divisor methods.

....

■ 5.3 DIVISOR METHODS OF APPORTIONMENT

In 1792, Hamilton's proposal was immediately met by a counter-proposal put forth by his chief political rival Thomas Jefferson. Jefferson's method (described below) seems to involve an enormous number of trial-and-error calculations, but this is not really true in practice. It is an example of a so-called "divisor method."

Jefferson's Method of Apportionment: Begin by choosing a whole number d (called a "divisor") as the desired size (population) of each "congressional district." Now allocate each state one seat for each congressional district. (That is, divide the state's population by the number d, and then round **down** to the nearest whole number to get that state's allocation.) If the number of seats allocated is exactly the House size, you're done. If it's less than the House size, go back and repeat the process with a larger choice for d. If it's more than the House size, use a smaller choice for d.

Politically, Jefferson's method won out, and it was used to apportion the House of Representatives for more than 50 years. However, by always rounding down, it systematically favors large states—a reduction from an ideal allotment of 49.9 to 49 leaves a state only about 2% short of ideal, whereas a reduction from an ideal allotment of 4.9 to 4 leaves a state about 20% short.

This favoritism of large states came to the forefront following the census of 1830 when John Quincy Adams, the Representative from Massachusetts, saw how badly the smaller New England states were faring in comparison with large states like New York. He proposed replacing Jefferson's method with one now known as Adams's method:

Adams's Method of Apportionment: Proceed exactly as in Jefferson's method except, where Jefferson rounds **down** to the nearest whole number, now round **up** to the nearest whole number.

Adams's method favors small states for exactly the same reason that Jefferson's method favors large states: a rounding up from 4.1 to 5 is a gain of roughly 20% while a rounding up from 49.1 to 50 is a gain of roughly only 2%. It was left to Daniel Webster to propose a more moderate alternative.

Webster's Method of Apportionment: Proceed exactly as in Adams's and Jefferson's methods except, where Jefferson rounds down and Adams rounds up, now simply round to the nearest whole number as one would normally do.

Historically, Webster's method went into effect following the 1840 census, but is was replaced a decade later by Hamilton's method, rediscovered by Samuel F. Vinton and often referred to as "Vinton's method." One might expect this to be the end of the story—at least for divisor methods—but there turns out to be one more natural way to round numbers, and the corresponding divisor method has been the one that has been in effect since the census of 1930. It uses the idea of the geometric mean.

DEFINITION. The *geometric mean* of two numbers A and B is the square root of the product AB. "Rounding according to the geometric mean" means, for example, that a number x between 4 and 5 gets rounded down to 4 if x is less than the geometric mean of 4 and 5 (i.e. $\sqrt{20}$), and rounded up otherwise. This rounds x down if $x < \sqrt{20}$ and up if $x > \sqrt{20}$.

The Hill-Huntington Method of Apportionment: Proceed exactly as in Adams's and Jefferson's and Webster's methods except, where Jefferson rounds down and Adams rounds up and Webster rounds in the normal fashion, now round according to the geometric mean.

The rationale behind the Hill-Huntington method can briefly be described as follows. In 1911, Joseph A. Hill, then the chief statistician in the census bureau, suggested a philosophical principle—as opposed to a method—on which apportionment should be based. He wanted to look at per capita representation, that is, a state's population divided by the number of seats. Thus, one state might have a per capita representation of 1,740,000 while another might have a per capita representation of 1,340,000 million. The difference, arrived at by subtracting, is 400,000. But what one really wants to look at here is the relative difference, in this case 400,000/1,340,000 = .2985 or 29.85%. It may happen that transferring a seat from the state with the smaller per capita representation to the one with the larger would reduce this relative difference, thus improving the equity. Hill's proposal was to find an apportionment method with the property that no two states could reduce the relative difference in per capita representation by such a transfer of a seat.

Edward V. Huntington, a professor of mathematics at Harvard, showed that the method now called the Hill-Huntington method does, in fact, satisfy Hill's principle.

■ 5.4 A GLIMPSE OF IMPOSSIBILITY

There is a remarkable result due to Michel L. Balinski and H. Peyton Young that the only apportionment methods that satisfy the population property are the divisor methods. But divisor methods, it turns out, are never guaranteed to satisfy the quota condition. Thus, we have a situation—an impossibility theorem—analogous to what we saw in the context of social choice.

While the Balinski-Young result is quite complicated, there is a weaker result that is nevertheless striking. It shows that a search for a perfect apportionment method is as doomed as the search for a perfect social choice procedure.

THEOREM. *There is no apportionment method that satisfies the monotonicity property, the quota condition, and the population property.*

PROOF. Assume that we have an apportionment method that satisfies the monotonicity property and the quota condition. We'll show that it must fail to satisfy the population property.

Consider the situation in which there are 7 seats, 4 states (A, B, C, and D), and a total population of 4200 distributed as follows: A has 3003, B has 400, C has 399, and D has 398. We can calculate the quota for each in the usual way (for example, A's quota is $[3003/4200] \times 7 = 5.005$). The results are as follows:

State	Population	Quota
A	3003	5.005
B	400	0.667
C	399	0.665
D	398	0.663

It is easy to see that because of the quota condition and monotonicity, the only possible apportionments are 5,1,1,0 and 6,1,0,0 (see Exercise 1). In particular, state A gets at least 5 seats and state D gets no seats.

Now suppose that at the next census there are 1100 additional people, with state A gaining 1, state D losing 1, and states B and D faring as show below:

State	Population	Quota
A	3004 (+1)	3.968
B	1503 (+1103)	1.985
C	396 (−3)	0.523
D	397 (−1)	0.524

Again, it is easy to see that because of the quota condition and monotonicity, the only possible apportionments are 4,2,0,1 and 4,1,1,1 and 3,2,1,1 (see Exercise 2). In particular, state A gets at most 4 seats and state D gets at least one.

Thus, state A has gained population and lost a seat, while state D has lost population and gained a seat. This completes the proof.

■ 5.5 DISPUTE RESOLUTION AND FAIR DIVISION

In Chapter 1, we studied social choice procedures, and evaluated each according to a set of reasonably fair criteria that we intuitively believe a social choice procedure should satisfy. In the last few sections, we looked at apportionment methods and again evaluated each according to a set of reasonably fair criteria that we intuitively believe an apportionment method should satisfy. Next we look at the issue of fairness in a different realm—dispute resolution. We will consider several methods of dispute resolution, and again evaluate each according to different notions of fairness.

Even most children are familiar with the method of "divide-and-choose." If two people want to fairly divide a candy bar in two pieces, one person will physically divide the candy, and the second person will choose which piece to take. Since the divider doesn't know which piece he will receive, it is clearly in his best interest to make the two pieces equal size, thereby guaranteeing that he will receive half. The chooser is also happy, of course, since she will definitely get the bigger piece. This method of division works equally well if the item to be divided is heterogeneous and the people's valuations of each piece differ. For example, we may be dividing a birthday cake into two pieces—one of us just wants as big a piece as possible, while the other would rather have a smaller piece if it contains more of the delicious frosting roses. The divider might split the cake into two pieces of unequal size, but if he values each piece equally, he will still be happy in the end since he is guaranteed a piece worth half the total value.

The divide-and-choose method described above is well known and commonly used. In fact, it has been used for thousands of years! In the ancient Greek text *Theogeny* written by Hesiod over 2700 years ago, Prometheus and Zeus divide some meat between them; Prometheus first splits the meat into two piles, and Zeus chooses a pile. A reference dating even further back can be found in the Hebrew bible in the story of Abram (later referred to as Abraham) and Lot. The two men travelled together, and eventually reached a piece of land that would not support both. To end the quarreling between the brothers' herdsmen, Abram proposed the following solution:

"Let there be no strife between you and me, between my herdsmen and yours, for we are kinsmen. Is not the whole land before you? Let us separate: if you go north, I will go south; and if you go south, I will go north." (Gen. 13:8-9)

Essentially, Abram divided the land, and Lot chose which piece he preferred.

The divide-and-choose method appears to be a great choice for two people who need to divide a single item, homogeneous or heterogeneous, but there are countless situations to which the method is not applicable. Perhaps more than one item needs to be divided, and some of the items are indivisible, like a TV. The TV will not be of much use if we cut it in two! It might also be the case that the item(s) needs to be split among three or more people. It is certainly not clear how to extend divide-and-choose to work with three people. In fact, the notion of fairness itself becomes more complicated when more than two people are involved. We will discuss fair division procedures for three or more people in Chapter 11.

In Chapter 1, we developed several properties that we used to evaluate the different social choice procedures. Similarly, we have several criteria to evaluate the fairness of a particular fair division procedure. Throughout our study of fair division, we will use the following criteria to evaluate the fairness of a given procedure.

DEFINITION. If there are n parties involved, we say that an allocation is *proportional* if each party receives at least $1/n$ of the total goods according to his own valuation. So with two people, an allocation is proportional if each party receives as least half the total value; with three people, an allocation is proportional if each party receives at least a third; and so on.

Note that it is possible (and occurs quite frequently in practice as we will later see) that each of the n parties receives strictly more than $1/n$ of the total. This may seem counterintuitive, and is, in fact, impossible if each party values the items exactly the same. But remember that each party typically values the items differently. For an extreme example, suppose that Annie and Ben are dividing a box of cookies that contains both peanut butter cookies and chocolate chip cookies. If Annie hates peanut butter and places no value on any of the peanut butter cookies, and Ben similarly detests chocolate, then the allocation in which Annie gets all the chocolate chip cookies and Ben receives all the peanut butter cookies leaves both parties with an impressive 100% of the total value!

We can think of the proportionality criteria as ensuring that each party receives a "fair share" of the goods. But how fair a share does this really guarantee? If Annie, Ben, and Chris are dividing a cake, and Annie receives 40% of the cake (according to her valuation), Ben receives 35% of the cake (according to his valuation), and Chris receives 50% of the cake (according to his valuation), then that allocation is proportional. If Annie values Ben's portion of cake at half the total value, however, then Annie would rather have Ben's piece than her own. Although the given allocation is proportional, Annie might not consider it fair. A stronger fairness criteria is the following:

DEFINITION. An allocation is *envy-free* if no party values someone else's portion of goods more than his own.

In the example in the last paragraph, the allocation is proportional, but not envy-free. As the following proposition shows, envy-freeness is generally a stronger criterion than proportionality, although the two are equivalent for two parties.

PROPOSITION. *If an allocation is envy-free, then it must also be proportional. For two parties, if an allocation is proportional, then it is also envy-free.*

PROOF. In order to show that every envy-free allocation is proportional, it suffices to show that if an allocation is not proportional, then it is not envy-free. For simplicity, we will assume that there are three people named *A*, *B*, and *C* (with the general case left for the exercises). Assume that we have an allocation that is not proportional and thus at least one of the people—assume it is *A*—thinks he has received less than one-third. Then A thinks that B and C between them have more than two-thirds. But this means that A thinks that at least one of B and C—assume it is B—has more than one-third. Thus, A thinks that he (A) has less than one-third and A thinks that B has more than one-third. Hence, A envies B, and so the allocation is not envy-free.

For the second part of the proposition, we assume that $n = 2$, so the allocation is between two people A and B. From the above paragraph, we know that if the allocation is envy-free, then is it proportional. On the other hand, if it's proportional, then each of A and B thinks he has at least one-half, and thus thinks that the other has at most one-half. But someone with at least one-half will never envy someone with at most one-half. Thus the allocation is envy-free, as desired.

Even envy-freeness, however, may not be enough to guarantee that all parties involved are happy with a given division of goods. For example, suppose that the following matrix gives the valuations of Annie, Ben, and Chris for a specific division of goods. The first column gives Annie's valuations of each of the three sets of goods, the second gives Ben's valuations, and the third gives Chris's valuations. Note that the numbers in each column must add to 100.

	Annie	Ben	Chris
Item 1	80%	30%	25%
Item 2	10%	40%	25%
Item 3	10%	30%	50%

Looking along the main diagonal, we see that the allocation in which Annie gets item 1, Ben gets item 2, and Chris gets item 3 is proportional—each number is strictly bigger than 33 1/3. The allocation is also envy-free since each of the entries on the main diagonal is greater than the other entries in that column. One can imagine though that Ben and Chris might not view this division as fair. Although neither would switch with Annie since both Ben and Chris value their lots as larger in value than hers, they might not be happy that she received 80% of the total goods in her opinion, whereas Ben and Chris received only 40% and 50% of the total in their own estimations. In a sense, Annie is happier with her portion than either Ben or Chris is with his.

DEFINITION. An allocation is *equitable* if each party involved receives exactly the same percentage of the total value according to his own valuations of the goods.

In the example given above, the allocation is envy-free (and therefore proportional), but not equitable. As we will see in the exercises, the criterion of equitability is independent from that of envy-freeness and proportionality. That is, there are allocations that are envy-free but not equitable and there are also allocations that are equitable, but not envy-free. The last criterion of fairness that we will consider here is that of efficiency, sometimes referred to in the literature as *Pareto-optimality*.

DEFINITION. An allocation is *efficient* if there is no other allocation possible that is at least as good for all parties and strictly better for at least one party.

In the cookie example above, the allocation in which Annie gets all the chocolate chip cookies and Ben gets all the peanut butter cookies is efficient. Both receive 100% of the total value, so neither can do better.

It is important to note that efficiency itself is not a good measure of fairness. For example, if one party receives everything, and all other parties receive nothing, then that allocation is efficient, since the parties receiving nothing cannot do better without the party who received everything doing worse. The allocation is hardly fair however! It is only in conjunction with the other criteria of proportionality, envy-freeness, and equitability, that efficiency is meaningful. Intuitively, efficiency ensures that there is no other allocation which would make everyone happier.

■ 5.6 AN ALTERNATIVE TO DIVIDE-AND-CHOOSE

We saw that divide-and-choose yields an envy-free division of a single heterogeneous good for two people. It is not equitable, however. The divider cuts the cake into two equal pieces and therefore receives a share worth exactly half the cake. Since the chooser typically does not value the two pieces equally though, he or she chooses the bigger piece and receives a share worth strictly more than half the cake. Austin's moving knife procedure, discovered in 1982 by A. K. Austin, gives an envy-free and equitable division for two people. The procedure works as follows.

Austin's Moving-Knife Procedure for Two People

Suppose Annie and Ben are dividing a rectangular cake between themselves.

Step 1: Annie places one knife at the left edge of the cake. She places a second knife at the dividing line, where she considers the pieces of cake on either side of the knife to be equally valuable.

Step 2: Annie then takes the two parallel knives and moves them across the cake so that at any instant, Annie believes the cake between the knives to be exactly half the cake.

Step 3: Ben yells "stop" when he considers the cake between the knives to be exactly half the cake. The cake is cut at the locations of the two knives.

Step 4: One person gets the piece of cake between the knives, and the other gets the two pieces on the outsides of the knives. It does not matter which person receives which portion.

If there is never a moment when Ben believes the cake between the knives to be equal to half the cake, then the procedure does not work since Step 3 never occurs. Fortunately, there must be such a moment for the following reason. When Annie originally places her knives over the cake Ben considers the cake to one side of the knife to be strictly less than half the cake. He must therefore consider the cake to the other side of the knife to be strictly more than half the cake. Let R denote this portion of cake. Now if Ben never yells "stop" as Annie moves the two knives, then the ending location of the knife on the left will be the original location of the knife on the right; the knife on the right is now at the rightmost edge of the cake. So the piece of cake between the knives is exactly R which Ben values at over half the cake. The knives originally bounded a piece of cake that Ben considered to be less than half the cake, and they end bounding a piece that Ben considers to be more than half the cake. The knives were moving continuously though, so at some moment the knives must have bounded exactly half the cake.

Since Ben yells "stop" only when he considers the cake between the knives to be exactly half, Ben sees the division as a 50-50 split. Similarly, since Annie kept the knives at whatever distance was necessary to ensure that the cake between the knives is half the cake, Annie also sees the division as a 50-50 split. Thus the procedure is equitable. It is also envy-free since each person sees each of the two portions of cake as equally valuable.

In the description above, we are implicitly assuming that Annie and Ben will act honestly, so we did not distinguish between the rules of the procedure and the strategy. If Ben is feeling greedy and hoping to benefit by misrepresenting his valuation of the cake, then he might not yell "stop" at the point where he sees both sides of the knife as equally valuable. In doing so, however, he risks getting a piece of cake worth less than half the total value. In general, when we describe a fair division procedure, we are describing the actions that each party should take to be guaranteed a particular outcome. Acting insincerely may, at times, result in a better outcome, but it may also result in a strictly worse outcome. We further discuss dishonest strategies in the exercises.

Unfortunately, Austin's method is not efficient. Suppose the cake is half vanilla and half chocolate. If Annie likes only vanilla and Ben likes only chocolate, then giving Annie the vanilla half and Ben the chocolate half gives each person a piece of cake worth 100% of the total value of the cake. The Austin procedure, however, will give Annie and Ben each exactly 50% of the cake's value. Since the first allocation is better for both, this shows that the Austin method is not efficient.

■ 5.7 ADJUSTED WINNER

Divide-and-choose and Austin's moving-knife procedure are two options for dividing a single good between two parties. Both are envy-free, only Austin's procedure is equitable, and neither is efficient (see Exercise 7). When there are several goods (possibly indivisible) or issues in dispute, however, we can find an allocation that is envy-free, equitable, and efficient, using the adjusted winner procedure. Adjusted winner is a method of dispute resolution for two parties that guarantees an outcome that is envy-free, equitable, and efficient. The procedure uses a point-allocation system, and requires only simple algebra to implement. Since the method is applicable when the dispute involves not only goods but also issues, we will refer to the *items* to be divided. For example, in a divorce settlement, a couple must often deal with custody arrangements as well as joint property. "Winning" the issue of custody might entail custody of the children on weekdays, while "losing" that item would mean weekend custody. For issues like these, the parties involved could determine together (or with a mediator) before the procedure is applied what constitutes winning and losing.

We will illustrate the procedure first with an example. Suppose that Annie and Ben are getting divorced, and the following items are under dispute.

1. House—The house is located very close to Annie's office, and Annie actually designed the recently renovated kitchen, so she values the house more than Ben does.

2. Investment Account—The investment accounts are Annie and Ben's combined life savings, and very valuable to both.

3. Baby Grand Piano—Although Annie has been taking piano lessons, Ben is the skilled pianist. It is his most prized possession.

4. Plasma TV—It was Ben's idea to purchase the Plasma TV, and he watches more TV than Annie does. He also uses it to screen many movies, and writes reviews for a local newspaper.

5. Tawny, the golden retriever—Tawny goes to work with Annie most days, so Annie spends much more time with Tawny than Ben does. She is very attached to her dog.

6. Car—Annie walks to work everyday, and rides her bike frequently, so the car is somewhat less valuable to Annie than to Ben.

Each party has 100 points to distribute over all the items according to which they value most. Annie and Ben's point distributions are below.

Annie	Item	Ben
35	House	15
20	Investments	25
10	Piano	25
5	TV	15
25	Dog	10
5	Car	10
100	Total	100

The division of items occurs in two stages. During the first stage, each item is initially awarded to the person who values it most. So Annie receives the house and the golden retriever, and Ben receives the investment account, baby grand piano, plasma TV, and the car. At this point, Annie has 60 points, and Ben has 75 points. Since Ben has more points, we say that Ben is the *initial winner*. The next stage is the equitability adjustment. We need to transfer items, or fractions thereof, from Ben to Annie until the point totals of each are equal and the allocation is thus equitable.

The order of the items to be transferred is important. To determine the order, we consider, for each of Ben's items, the ratio of the points assigned by Ben to the item to the points assigned by Annie to the item. Note that each of these ratios will be at least 1, since Ben received the items to which he had assigned more points. In the example at hand, the ratios for each of Ben's items are as follows:

$$\text{Investments}: \frac{25}{20} = 1.25$$

$$\text{Piano}: \frac{25}{10} = 2.5$$

$$\text{TV}: \frac{15}{5} = 3$$

$$\text{Car}: \frac{10}{5} = 2$$

The transfer of items starts with the item for which the ratio above is the smallest, then the next-smallest, and so on. Intuitively, this is the fairest way to proceed since the "cost" to Ben per point transferred to Annie is smallest. For example, transferring the TV requires lowering Ben's point total by 3 points for every 1 point transferred to Annie, while transferring the car would only lower Ben's point total by 2 for every 1 point transferred to Annie. We will see in Chapter 11 that this order is crucial to the proof that the resulting allocation is efficient.

Since the ratio for the investments is the smallest, we begin with that. Notice that if we were to transfer the entire investment portfolio to Annie, then Annie would have more points than Ben. A simple algebraic calculation will give the exact fraction of the investments to be transferred in order to achieve equitability. Let x be the fraction of the investments transferred to Annie, so that $1 - x$ is the fraction retained by Ben. After the transfer, then, Annie will have 60 points (from the house and dog) plus $20x$ (her portion of the investments), while Ben will have 50 points (from the piano, TV, and car) plus $25(1-x)$ (his portion of the investments). To guarantee that the resulting point totals are equal, we need to ensure that

$$60 + 20x = 50 + 25(1 - x) = 75 - 25x.$$

Thus

$$45x = 15,$$

so $x = 1/3$. In the end, then, Annie receives the house, the golden retriever Tawny, and one-third of the investment portfolio, while Ben keeps the piano, TV, car, and two-thirds of the investments. Each person walks away with an impressive total of 66 2/3 points, well over half the total value.

In this example, splitting up the investment portfolio is not a difficult task, at least for the stock brokers. If we had needed to split the piano, however, it certainly wouldn't be simple since a third of a piano is not very valuable to anyone! In this case, a mediator might perform the adjusted winner procedure for Annie and Ben, and then reveal that the piano needs to be split: one person will receive one-third, the other two-thirds (without divulging who receives which portion). Together, then, Annie and Ben might decide to sell the piano and split the profits according to the prescribed proportions. Or they might decide that if Annie receives the larger half, they will sell the piano, but if Ben receives the larger share, he will buy out Annie's share. If the golden retriever is to be divided, they might decide to share custody. Many options are available when splitting items becomes necessary; fortunately, as we will see, at most one item need ever be split.

Now that we've seen an example, let's look at the adjusted winner procedure in general.

Adjusted Winner Procedure

Step 1: Each item, for which there is no tie in point values, is initially awarded to the party who awarded it more points. Next, in any order and one at a time, the tied items are given to whomever has fewer points at the time. If the point totals are equal, then a tied item can be given to either party.

Step 2: If the point totals of each party are equal at the end of Step 1, then the procedure is done; an equitable allocation has been achieved. Otherwise, the equitability adjustment (Step 3) occurs.

Step 3: Call the party with more points at the end of Step 1 the initial winner. Calculate the ratio, for each item awarded to the initial winner during Step 1, of the points awarded to the item by the initial winner to the points awarded to the item by the other party. In order of ascending ratios, transfer items (or fractions thereof) from the initial winner to the other party until the point totals are equal.

Note that the procedure does in fact end. Only finitely many items are under dispute. If after an item is transferred, the initial winner still has the larger point total, then the next item is transferred. If transferring an item results in equal point totals, then the procedure is finished. If transferring an item would result in the initial winner having fewer points than the other party, then that item must be split; the procedure is then finished. Also note that the procedure can be modified in the case of unequal entitlements, for instance if a prenuptial agreement indicated that the shared property be divided 60% - 40%. See Exercise 15 for a specific example.

The adjusted winner procedure is widely applicable to many types of disputes, and is easy to use without hiring an expensive expert negotiator. A mediator could be useful, however, in performing the procedure and helping the parties to both identify issues at dispute and identify what winning at each issue entails. Allocating points to the items at hand is not necessarily an easy task, but it does give each party a degree of control in a difficult situation. Moreover, determining one's valuations can be done alone, with no time pressure, and can be much less stressful than worrying about an emotional battle of wits with a skillful debater. Better yet, if both parties submit honest valuations, then each is guaranteed an outcome that is envy-free, equitable, and efficient!

Equitability. It is easy to see that the procedure is equitable by design. The procedure ends when the point totals of each party are equal.

Efficiency. Efficiency is often the most difficult of the three criteria to satisfy, and therefore one of the most remarkable properties of the adjusted winner procedure. The proof of efficiency, given in Chapter 11, is one of the more complicated proofs that we present in the book.

Envy-freeness. This property follows, in fact, from the other two when exactly two parties are involved. Suppose, for contradiction, that the allocation is equitable and efficient, but not envy-free. Since envy-freeness and proportionality are equivalent for two parties, then it must be the case that at least one of the parties received less than half according to his own valuation. But equitability then implies that both parties received less than half. This allocation is not efficient because we can find another division in which both players do better: give each party's share to the other party. If each party originally received x points, where $x < 50$, then now each receives $100 - x > 50$ points, so this allocation is strictly better for both parties involved, contradicting the efficiency of the original division.

Although adjusted winner can be a great method of dispute resolution, it is clearly not always applicable. Sometimes disputes involve more than two parties, and in Chapter 11, we will look at some alternative methods that can be used for more than two parties. Unfortunately, just as there is no perfect social choice procedure and no perfect method of apportionment, there is no perfect fair division procedure for three or more parties. We will see that it is impossible to guarantee an envy-free, equitable, and efficient division for more than two people!

■ 5.8 ADJUSTED WINNER AND THE MIDDLE EAST

Most of the examples we have looked at so far have been concerned with dividing a number of physical goods, but the adjusted winner procedure can also be applied to issues in all sorts of disputes. Here we illustrate with an example applying the adjusted winner procedure to the Israeli-Palestinian conflict in the Middle East. We will give a simplified version addressing only 5 key areas of disagreement between the Israelis and Palestinians. For a more in-depth treatment, see T. G. Massoud's paper in the *Journal of Conflict Resolution* (June 2000) which considers nine key issues of disagreement. The following five issues are some of the most contentious sources of dispute between the Israelis and Palestinians.

1. **West Bank:** Several areas of the West Bank are inhabited by Israelis who have no desire to leave their homes. The Palestinians, however, believe that these settlements are illegal, and that the Israelis should evacuate.

2. **East Jerusalem:** In 1967, Israel unified control over all of Jerusalem by defeating Jordanian forces in the Six Days War. A majority of the residents of East Jerusalem are Palestinian, however, and both Israelis and Palestinians argue that East Jerusalem is central to their sovereignty.

3. **Palestinian Refugees:** Israel has refused to recognize that its establishment and expansion in 1948 and 1967 displaced Palestinian villages and communities. The Palestinians insist that Israel recognize the refugees' "right of return" to Israel, and provide compensation for the refugees and to Arab states that have hosted the refugees.

4. **Palestinian Sovereignty:** Israel does not recognize Palestine as a sovereign nation.

5. **Security:** There are several security issues involved in the Israeli-Palestinian conflict. Some Israelis fear that terrorism would flourish under a Palestinian state that lacks the means to effectively fight terrorism. Specific security issues include: border control, control of airspace, security in Jerusalem, and "early warning stations" in the West Bank and Gaza that would assuage Israeli concerns against surprise attacks but provide insufficient military capability to threaten Palestinian forces.

It is, of course, impossible to know exactly how Israeli and Palestinian leaders would allocate points to the issues above, and moreover, there would be widespread disagreement as well among each nation's inhabitants. Massoud, a political scientist at Bucknell University, examined expert opinions, interim agreements, and working plans to arrive at a reasonable approximation of possible point allocations by each side. His research (modified for this simplified version of the dispute) suggests that one possible point allocation is as follows.

Israel	Issue	Palestine
22	West Bank	21
25	East Jerusalem	23
12	Palestinian Refugees	18
15	Palestinian Sovereignty	24
26	Security	14
100	Total	100

In the first stage of the adjusted winner procedure, Israel wins the issues of the West Bank, East Jerusalem, and security, while Palestine wins the issues of refugees and sovereignty. At this point, Israel has 73 points and Palestine has 42 points. Since Israel is the initial winner, then we look at the ratios of points for the issues won by Israel:

$$\text{West Bank}: \frac{22}{21}$$

$$\text{East Jerusalem}: \frac{25}{23}$$

$$\text{Security}: \frac{26}{14}$$

The equitability adjustment begins with the West Bank since $22/21 < 25/23 < 26/14$. Transferring the entire issue would give the Palestinians more points than the Israelis. To determine the percentage of the issue of the West Bank each party receives, we solve for x in the following equation.

$$51 + 22x = 42 + 21(1 - x) = 63 - 21x$$

$$43x = 12$$

$$x = 12/43 \approx 2/7$$

Thus the Israelis are left with the issues of East Jerusalem, security, and roughly 2/7 of the issue of the West Bank. The Palestinians are left with the issues of refugees, sovereignty, and roughly 5/7 of the issue of the West Bank. Splitting the issue of the West Bank according to the prescribed proportions might be as simple as giving 2/7 of the land to the Israelis and 5/7 to the Palestinians.

■ 5.9 CONCLUSIONS

We began this chapter by considering the issue of apportionment. Just as we saw in Chapter 1 with social choice procedures, there are several different methods for apportionment that seem reasonable, so we introduced three criteria by which we could measure the "fairness" of each procedure: the monotonicity property, the quota property, and the population property. Unfortunately, we saw that it is impossible to find a single apportionment method that satisfies each of the three criteria.

Next, we turned to the issue of fairness in dispute resolution. We looked at the divide-and-choose method for dividing a single good between two parties, but just as we saw for social choice procedures, the issue of dividing goods becomes considerably more complicated when three or more parties are involved. Our strategy of analysis, however, remained the same: we considered several different criteria by which we could measure the fairness of the fair division procedures. We will revisit these criteria in Chapter 11 when we look at fair division procedures for three or more parties.

The rest of the chapter was devoted to adjusted winner, a procedure that guarantees each of two parties an efficient, envy-free, and equitable allocation of goods or issues. We looked at an example of how the procedure could be applied to the Israeli-Palestinian conflict in the Middle East.

EXERCISES

1. Prove that in the proof of the impossibility theorem in **Section 5.4**, the quota condition and monotonicity imply that the only possible apportionments in the first example are 5,1,1,0 and 6,1,0,0.

2. Prove that in the proof of the impossibility theorem in **Section 5.4**, the quota condition and monotonicity imply that the only possible apportionments in the second example are 4,2,0,1, 4,1,1,1, and 3,2,1,1.

3. Use the Hamilton method to round each of the following numbers in the sum to a whole number, preserving the fact that the total is 20:

$$2.71 + 3.49 + 0.64 + 2.07 + 9.51 + 1.58 = 20$$

4. Apportionment methods can also be used in non-political contexts. Consider, for example, the situation in which a mathematics department has 10 faculty members, each of whom will teach 2 classes in the fall semester. These 20 sections need to accommodate seven different courses with enrollments (and "quota" to be commented upon momentarily) as follows.

Course	Enrollment	Quota
Calculus I	121	5.45
Calculus II	94	4.23
Calculus III	76	3.42
Linear Algebra	48	2.16
Real Variables	20	0.91
Cryptology	24	1.08
Math and Politics	61	2.75
Total	444	20

(a) Explain how the quotas were calculated.

(b) Find the apportionment of the 20 sections among the 7 courses resulting from the use of Hamilton's method.

5. Prove that for any number of parties, if an allocation is envy-free, then it must also be proportional.

6. Prove that if a division of goods between two parties is equitable, then one of the following is also true.

   (i) The division is also envy-free.

   (ii) If the two parties switch shares, then the division is envy-free.

7. (a) Prove that the divide-and-choose procedure does not guarantee an efficient allocation.

   (b) Prove that Austin's procedure does not guarantee an efficient allocation.

8. (a) Describe a situation in which you would rather be the chooser in the divide-and-choose procedure.

   (b) Describe a situation in which you would rather be the divider in the divide-and-choose procedure.

9. Is the divide-and-choose procedure manipulable? That is, can one player achieve a strictly better outcome by misrepresenting his true valuation of the cake? Prove that it is not, or give an example where one player achieves a strictly better outcome than obtained with an honest application of the divide-and-choose procedure.

10. Suppose that Annie and Ben are getting a divorce. The items under dispute are: the lease to their apartment, a one-year old blue MINI-Cooper, a baby-grand piano, a plasma TV, a sailboat, and a Golden Retriever named Tsuki. They assign the following points to each item:

Annie	Item	Ben
35	Apartment	30
20	Mini-Cooper	20
15	Piano	20
5	TV	10
5	Sailboat	10
20	Dog	10
100	Total	100

Determine what items Annie and Ben each receive according to the adjusted winner procedure.

11. (a) Determine the allocation determined by the adjusted winner procedure for the following example.

Ross's Points	Item	Rachel's Points
35	Manhattan Apartment	30
50	Custody of Daughter Emma	50
10	Share in ownership of local coffee shop	15
5	Right to spend Thanksgiving with Monica and Chandler	5
100	Total	100

   (b) Explain why, in this example, each party gets relatively few points overall.

12. Katie and Hubbell Gardiner are getting a divorce. They have valued the items to be divided as follows. Determine who gets what under the adjusted winner procedure.

Katie's Points	Item	Hubbell's Points
25	Hubbell's first story: "A Country Made of Ice Cream"	15
10	Hollywood Home	35
25	Custody of Daughter	15
15	Hubbell's Navy Uniform	15
10	Katie's Hair Iron	10
15	Typewriter	10
100	Total	100

13. Allie and Michael are getting a divorce. The five major items to be divided are: custody of Cauchy the cat, the wine collection, season passes to the New York City Ballet, a pair of tap shoes autographed by Savion Glover, and a first edition copy of *Walden* by Thoreau. Their point allocations are as follows. Determine who gets what according to the adjusted winner procedure.

Allie's Points	Item	Michael's Points
20	Cat	40
10	Wine Collection	5
25	Ballet Tickets	25
30	Tap Shoes	20
15	Book	10
100	Total	100

14. Emma and Kate are planning to open a new restaurant, and have several projects to finish before they will be ready to open. They would rather split up the projects between them so that each person has full control of a few specific issues instead of working together on each of the different projects. Each person has devoted 100 points to the projects listed below. Use adjusted winner to determine who will be in charge of which issues. If one project is to be divided, outline a possible division of labor for the specific project to be divided.

Emma's Points	Item	Kate's Points
20	Menu Design	10
25	Interior Design	15
10	Advertising	5
15	Dining Room Layout	20
10	Bar Layout	20
10	Hiring Waitstaff	15
10	Hiring Chefs	15
100	Total	100

15. The adjusted winner procedure can be adapted for unequal entitlements. Suppose, for example, that Annie and Ben are getting a

divorce, but they signed a pre-nuptial agreement that gives Annie 60% of the joint property and Ben 40%. During the "equitability" adjustment stage of the adjusted winner procedure, Annie's point total should be exactly 1.5 times that of Ben. Determine the allocation dictated by adjusted winner for the following items.

Annie's Points	Item	Ben's Points
35	Right to retain lease on apartment	30
20	Entertainment System	15
15	Pool table	25
15	Antique Table	20
15	Washer & Dryer	10
100	Total	100

16. Another method for dividing goods or issues between two people is *balanced alternation*, wherein the two parties take turns choosing issues and the party that chooses second is compensated by being able to choose two items during his first turn. For example, if persons A and B are dividing six goods between them, then they might choose in the following order: A, B, B, A, B, A.

   (a) If Emma and Kate, in Exercise 14, use balanced alternation rather than adjusted winner, what is the final allocation of issues?

   (b) Is Emma better or worse off with balanced alternation than with adjusted winner? What about Kate?

   (c) Describe a particular example of two sets of goods to be divided where adjusted winner is far better than balanced alternation.

   (d) Is there any situation in which the parties might prefer to use balanced alternation over adjusted winner?

CHAPTER

6

Escalation

■ 6.1 INTRODUCTION

Important examples of escalation are easy to find in political science, such as the buildup of American troops in Vietnam during the 1960s and the arms race of the 1960s, 1970s, and 1980s, to mention just two. Such escalatory behavior is driven at least in part by a desire to keep previous investments from having been wasted. In this chapter we consider a model of escalatory behavior introduced by the economist Martin Shubik and extensively analyzed by Barry O'Neill. This model is known as the dollar auction.

Regular auctions do not elicit escalatory behavior since one can always quit and be back where one started. In the dollar auction, however, the rules are changed so that when the bidding for the prize (typically a dollar—hence the name) is completed, both the highest and the second-highest bidder pay the auctioneer what he or she bid, although only the highest bidder gets the prize. Hence, it is worse to have bid and lost than never to have bid at all. An appreciation of the escalatory nature of the dollar auction is certainly enhanced by the opportunity to watch a group of one's (unsuspecting) peers actually involved in the bidding process. Remarkable though it may seem, winning bids typically exceed one dollar, with the auctioneer often pocketing more than three dollars, and having to give out only one dollar.

The set-up for the dollar auction is given in **Section 6.2**, while in **Section 6.3** we provide an analysis of "rational behavior" in the dollar auction for some fairly simple cases. The method of analysis used is an important one; it is the naïve, straightforward, "brute-force" attack that involves the organized presentation of all possibilities in such a way that optimal strategies can be methodically—if not easily—identified. A similar analysis was done with the chair's paradox (**Section 3.7**) and the theory of moves (**Section 4.7**).

In **Section 6.4** we introduce the idea of a "back-of-the-envelope calculation" and we demonstrate some serious limitations to the kind of analysis done in **Section 6.3**, even in the presence of high-speed computers. **Section 6.5** deals with a beautiful theorem due to Barry O'Neill that provides a complete analysis of the dollar auction without ever looking at the associated game tree. The statement of the theorem is easy to understand. The proof of the theorem is more difficult, although it requires absolutely no mathematical preliminaries. It can be found in Chapter 12. We conclude in **Section 6.6** with a treatment of Vickrey auctions.

■ 6.2 THE DOLLAR AUCTION

The setup for the dollar auction is as follows. We assume there are only two people bidding, one of whom is designated to go first. The one going first must bid. That is, passing is not an option at this point. A given unit of money is fixed (for example, a unit may be a dime), and no bid can involve a fractional part of a unit. We let s denote the number of units in the stakes. (So if the units are dimes and the stakes one dollar, then $s = 10$.) We also assume that each bidder has a fixed number of units of money at his or her disposal to bid with and we assume it is the same number for both bidders. Let b (for "bankroll") denote this number; e.g., if units are again dimes, then $b = 120$ corresponds to both bidders having a bankroll of twelve dollars. Bidding proceeds as usual with no communication, deals, threats, etc. allowed. When the bidding is concluded (and the fact that the bankroll is fixed guarantees that it definitely will be concluded at some point), the higher bidder pays what he or she bid and receives the stakes s; the other bidder pays what he or she bid and receives nothing.

Following O'Neill, we will assume that both bidders in the dollar auction adhere to the *conservative convention*: if either bidder determines that two or more bids will lead to the same eventual outcome for himself or herself (and no bid leads to a better outcome), then that bidder will choose to make the conservative (i.e., smaller) bid, where we regard a pass as a bid of zero. For example, if $b = s$ and the first bidder opens with a bid of $b - 1$, then passing and bidding b yield the same eventual outcome—money neither lost nor gained—for the second bidder. Hence, according to the conservative convention, the second bidder would pass in this situation.

The dollar auction is both interesting to think about and entertaining to watch. From the point of view of political science, in fact, its primary usefulness is as an observable model of escalatory behavior in people. This "observable model" is what O'Neill refers to as the "real dollar auction" in order to distinguish it from what he calls the "ideal dollar auction." For most of this chapter, however, our concern will be with this "ideal" version of the dollar auction where one assumes complete rational analysis and rational behavior on the part of both players. The benefits of this game-theoretic analysis of rational bidding in the dollar auction are far less in terms of a deeper understanding of the dollar auction model itself than in an introduction to the (mathematical) ideas of analyzing decision trees and strategies.

■ 6.3 GAME-TREE ANALYSES

The results obtained here and in the next section will show that optimal strategies in this escalatory model are far from obvious. For example, if the units are dimes, the stakes one dollar (i.e., $s = 10$), and the bankroll twelve dollars (i.e., $b = 120$), then the optimal opening bid is thirty cents (i.e., three units). The reader should find this—at the moment—enormously nonobvious.

The situation that will be considered here is the one where both the bankroll b and the stakes s are three units. The "organized presentation of all possibilities" referred to above will involve a tree whose nodes

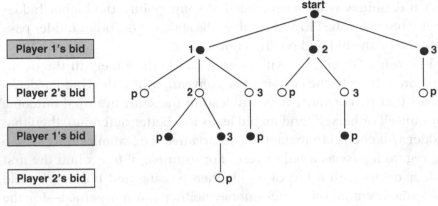

Player 1's bid	
Player 2's bid	
Player 1's bid	
Player 2's bid	

FIGURE 1

(the black and white dots in Figure 1 above) are labeled by numbers indicating a legitimate bid possibility for either the first bidder (who we'll call Player 1) or the second bidder (Player 2). All the nodes in a single horizontal row correspond to bids by the same player, and that player is identified to the left of that row.

Again, the node at the top labeled "start" has no particular significance except to make the tree appear connected instead of disjointed. The three nodes on the level below the start node represent the three possible opening bids for Player 1: one unit, two units, or three units. Since $b = 3$, no higher bid is possible. The three nodes on the level below the node on the left labeled "1" and connected to it by lines represent the responses that Player 2 can make to an opening bid of one unit by Player 1. And so on.

The nodes labeled "p" (for "pass") on this tree of possibilities are called *terminal nodes* (terminal in the sense of being final—no further bidding takes place). Any sequence of nodes beginning with "start" and moving down the tree—but never up—until it reaches one of these terminal nodes is called a *branch* of the tree. In particular, there are seven branches in this tree corresponding to the seven terminal nodes labeled with a p. Notice that these seven branches correspond to the seven possible bidding scenarios that can take place in the dollar auction where the bankroll is three units. Notice also that the fact that the stakes s is three units has not yet played any role.

The next thing we want to do is to embellish our "tree of possible bidding scenarios" with labels that indicate the loss or gain for each

player in each of the seven scenarios. That is, we will label each of
the seven terminal nodes with an ordered pair (i, j) indicating that the
result of following that bidding scenario (i.e., that branch) will be a
gain of i units for Player 1 and a gain of j units for Player 2. Both can't
actually gain, so a label such as $(-2, 0)$ means that Player 1 really loses
two units while Player 2 breaks even (see Figure 2 below).

To explain just one of these labels, notice that the lowest p on the
tree is labeled $(0, -2)$. The bidding scenario here is an opening bid of
one unit by Player 1, a response of two units by Player 2, and a "close-
out" bid of three units by Player 1. (It is a "close-out" in the sense that
the bankroll b is three units so Player 2 is now forced to pass, ending
the bidding.) Thus, Player 1 pays the three units bid and receives the
stakes s, which is also three units. Hence, Player 1's gain is zero. Player
2, of course, loses the two units he or she bid and this explains the "-2"
occurring in the label "$(0, -2)$."

Before continuing our analysis, we must tackle the question of
exactly what is meant by "rational play." Part of what it means is that
each player will bid in such a way as to maximize his or her gain (or
minimize his or her loss), as opposed to, for example, trying to maximize the difference between what he or she gains and that of the other
player. But it means more than this. It also means that each player is
aware that the other is operating this way. Hence, in deciding how to
maximize one's gain, one can look ahead and know that one's opponent will respond in a way consistent with maximizing his or her own

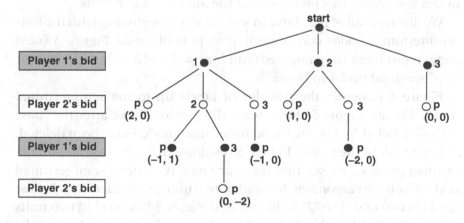

FIGURE 2

gain. It doesn't quite stop there. Rational play also means that each is aware that the other is aware that he or she operates in this way, and so on. We also assume that not only does each player follow our convention about choosing the more conservative of two equivalent bids, but that each knows the other is following this convention, and each knows the other knows that he or she is following this convention, and so on.

We are now ready to begin the process of analyzing rational play for this auction. The procedure used is again backward induction. It involves deleting terminal nodes and transferring the outcome labels to a node one level higher, and then repeating this process until the analysis is complete.

For the sake of describing this pruning process, let's call a node *semiterminal* if it is not a terminal node and all the nodes on the next level that are connected to it have outcome labels. For example, in Figure 2 the semi-terminal nodes are precisely the four nodes corresponding to a bid of three. Given our rationality assumption (and our conservative convention), it is easy to see exactly what outcome will result if the bidding scenario reaches a semi-terminal node: the player whose turn it is will simply choose the (least) bid corresponding to a next lower node with the best value for himself or herself. Our procedure will be to transfer the appropriate outcome label from a terminal node to a semi-terminal node, and then delete all of the terminal nodes that follow semi-terminal nodes. These semi-terminal nodes then become terminal nodes and so we can again search for (new) semi-terminal nodes and repeat this process until the analysis is complete.

We illustrate all of the above in several steps, beginning with the four semi-terminal nodes corresponding to bids of three. Figure 3 (next page) reproduces the game tree (still for $s = 3$ and $b = 3$) with the four semi-terminal nodes in boxes.

Figure 4 indicates the transfer of labels up to the semi-terminal nodes. Finally, Figure 5 shows what the tree looks like after the "pass nodes" (labeled "p") below the semi-terminal nodes have been deleted.

If we look at the tree in Figure 5 resulting from the first step in the pruning process, we see that there are now two (new) semi-terminal nodes, both corresponding to a bid of two units. Consider the leftmost (and lowest) one. The "2" indicates that Player 2 has just bid two units and so Player 1 has the choice of passing—and getting $(-1, 1)$ as the

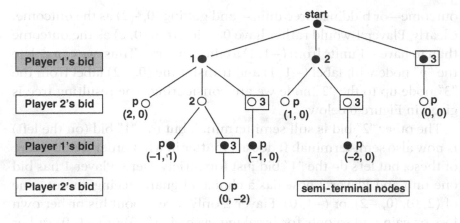

FIGURE 3

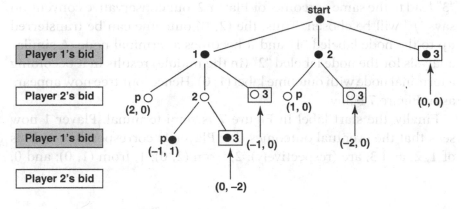

FIGURE 4

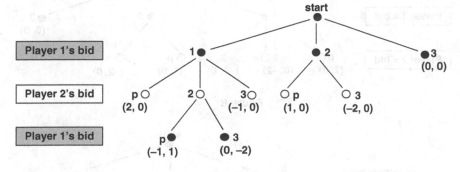

FIGURE 5

outcome—or bidding three units—and getting (0, −2) as the outcome. Clearly, Player 1 would rather have 0 units from (0, 2) as the outcome than to have −1 units from (−1, 1) as the outcome. Thus, we can delete the "p" node with label (−1, 1) and transfer the (0, −2) label from the "3" node up to the "2" node we are considering. The resulting tree is given in Figure 6 below.

The other "2" bid is still semi-terminal, but the "1" bid (on the left) is now also semi-terminal. It doesn't matter in what order we take care of these, but let's do the "1" bid just for variety. Here, Player 1 has bid one unit and Player 2 now has a choice of guaranteeing an outcome of (2, 0), (0, −2), or (−1, 0). Player 2 only cares about his or her own loss or gain and so opts for breaking even via (2, 0) or (−1, 0) rather than losing two units via the outcome (0, −2). Since bids of "p" and "3" lead to the same outcome for Player 2, our conservative convention says "p" will be chosen. Thus, the (2, 0) outcome can be transferred up to the node labeled "1" and it becomes a terminal node. A similar analysis for the node labeled "2" (in the middle) results in it becoming a terminal node with outcome label (1, 0). Hence, our tree now appears as in Figure 7 below.

Finally, the start label in Figure 7 is semi-terminal. Player 1 now sees that the eventual outcomes (for Player 1) corresponding to bids of 1, 2, and 3, are (respectively): 2, from (2, 0); 1, from (1, 0); and 0,

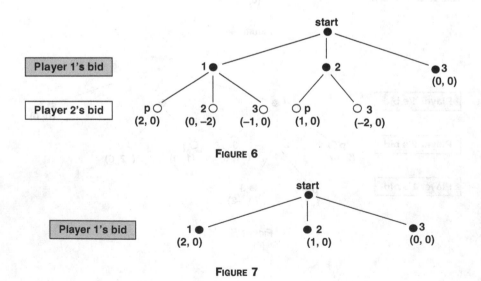

FIGURE 6

FIGURE 7

from (0, 0). Clearly 2 as an outcome is preferred so Player 1's opening bid will be one unit. Looking back to Figure 6, we can also see that Player 2 will now pass. Thus, the optimal strategies are for Player 1 to bid one unit and for Player 2 to pass. This completes the game-tree analysis for the dollar auction in the special case where both the stakes s and the bankroll b are three units.

In the above presentation of the pruning process, notice that we continually redrew the game tree as we proceeded. Moreover, we actually deleted, from the picture, those nodes that we no longer needed to consider. There are at least two drawbacks to this notational way of proceeding: the redrawing of the tree is tedious and the final result (i.e., the final tree) gives no "history" of the pruning process. A better notation, once the pruning process is understood, is simply to cross out the edges that we no longer need to consider, and to rewrite the labels as they get moved up in the tree. Our crossing-out will leave a pattern that looks like railroad tracks as in the theory of moves in Chapter 4. This is illustrated in Figure 8 below.

When the bankroll exceeds 3, the reader may want to consider deleting the pass nodes that occur after a "closeout bid" of b. This, for example, will give a tree as in Figure 5 to start the pruning process. The gain in working with the smaller tree probably justifies carrying a small piece of the process mentally.

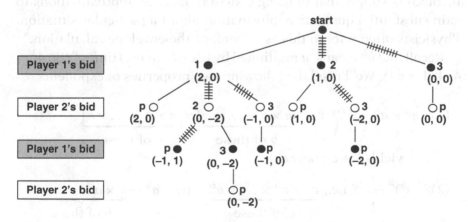

FIGURE 8

■ 6.4 LIMITATIONS AND BACK-OF-THE-ENVELOPE CALCULATIONS

In the previous section we saw how game trees provide a notation that permits one to analyze a dollar auction by explicitly looking at every possible scenario. This is an example of a very fundamental problem-solving technique: roll up your sleeves and look at all cases. On the other hand, the particular game tree we considered in the previous section corresponded to noticeably small values for the bankroll ($b = 3$) and stakes ($s = 3$). If we consider the kind of dollar auction often done for illustrative purposes in class where, say, the units are dimes, stakes are one dollar ($s = 10$), and the bankroll twelve dollars ($b = 120$), then we are clearly talking about an unwieldy tree with which to deal. If this were the first half of the twentieth century, probably nothing more regarding limitations of the game-tree analyses would need be said. But now, of course, we have computers. So the question we address here is the following:

Given a state-of-the-art computer, how long would it take you to do the game-tree analysis for $s = 10$ and $b = 120$?

The answer will turn out to be a well-known expression and our point about the limitations of this brute-force technique will have been made. There is, however, an additional benefit to be derived from the process undertaken in answering the above question. The idea involved is simply that of using coarse numerical approximations to gain substantial qualitative information about a particular situation. Physicists often refer to this as a "back-of-the-envelope calculation."

Recall that a^b means "a multiplied by itself b (many) times." (So $4^3 = 4 \times 4 \times 4 = 64$). We'll need the following two properties of exponents:

(1) $a^b \times a^c = a^{b+c}$ i.e., $\underbrace{[a \times a \times \ldots \times a]}_{b \text{ of these}} \times \underbrace{[a \times a \times \ldots \times a]}_{c \text{ of these}}$

yields $b + c$ many as)

(2) $(a^b)^c = a^{bc}$ i.e., $\underbrace{a^b \times a^b \times \ldots \times a^b}_{c \text{ of these}}$ where $a^b = \underbrace{a \times a \times \ldots \times a}_{b \text{ of these}}$

yields

$$\underbrace{\overbrace{a \times a \times \ldots \times a}^{b \text{ of these}} \times \overbrace{a \times a \times a \times \ldots \times a}^{b \text{ of these}} \times \ldots \times \overbrace{a \times a \times \ldots \times a}^{b \text{ of these}}}_{c \text{ of these}}$$

so we have $b \times c$ "as" all together.

The real key to the limitations that will arise involves the size of the game tree. In particular, our computer analysis will (at least) involve having each terminal node looked at (i.e., considered). So, how many terminal nodes are there? Notice that this depends on the size of the bankroll b, but not on the stakes s. For $b = 1$, 2, and 3, the trees are shown in Figure 9 below.

So, for $b = 1$, there is one terminal node; for $b = 2$, there are three terminal nodes, and for $b = 3$ there are seven. In general, it turns out that there are $2^b - 1$ terminal nodes for bankroll b. (If $b = 1$, then $2^b - 1 = 2^1 - 1 = 2 - 1 = 1$; if $b = 2$, then $2^b - 1 = 2^2 - 1 = 4 - 1 = 3$; if $b = 3$, then $2^b - 1 = 2^3 - 1 = 8 - 1 = 7$.) Hence, for our bankroll of twelve dollars, where the units are dimes (so we have $b = 120$), the computer must do at least 2^{120} things; i.e., it must look at $2^{120} - 1$ terminal nodes plus print an answer.

Now, using our properties of exponents, we see that:

$$2^{120} = 2^{4 \times 30} = (2^4)^{30} = 16^{30} > 10^{30}.$$

Hence the computer must do more than 10^{30} things.

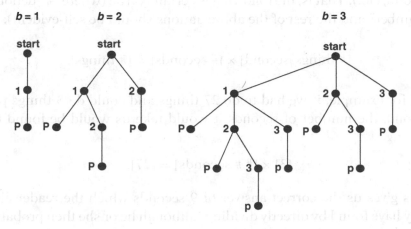

FIGURE 9

Let's assume that a state-of-the-art computer can do at most a quadrillion things per second. Note that since a quadrillion is a thousand trillion, and a trillion is a thousand billion, and a billion is a thousand million we have

$$1 \text{ quadrillion}$$
$$= 1,000 \times 1 \text{ trillion}$$
$$= 1,000 \times 1,000 \times 1 \text{ billion}$$
$$= 1,000 \times 1,000 \times 1,000 \times 1 \text{ million}$$
$$= 1,000 \times 1,000 \times 1,000 \times 1,000,000$$
$$= 10^3 \times 10^3 \times 10^3 \times 10^6$$
$$= 10^{15}.$$

Hence, the computer can do at most 10^{15} things per second.

Now that we know the computer must do (roughly) 10^{30} things and that it can do (roughly) 10^{15} things per second, we can ask how many seconds it will take to complete the task. For some readers, the answer may quickly suggest itself. Nevertheless, we shall belabor one reasonable line of thought leading to the answer, since a variant of this line of thought will occur twice more before the end of this section.

The line of thought we shall pursue is the one that makes use of the commonsense guidance provided by the units (seconds, things per second, etc.). That is, the following is certainly true (where "#" denotes "number" and the rest of the abbreviations should be self-evident):

$$[\# \text{ things/second}] \times [\# \text{ seconds}] = [\# \text{ things}].$$

So, for example, if we had to do 27 things and could do 3 things per second, the number of seconds it would take us would be found by solving:

$$[3] \times [\# \# \text{ seconds}] = [27].$$

This gives us the correct answer of 9 seconds which the reader also may have found by directly dividing (although he or she then probably checked this answer by calculating $3 \times 9 = 27$ with the units in mind).

If we go back to our computer calculation, then our equations (in units and numbers) are:

$$[\# \text{ things/second}] \times [\# \text{ seconds}] = [\# \text{ things}]$$

and

$$[10^{15}] \times [\# \text{ seconds}] = [10^{30}].$$

So what is [# seconds]? Think of it as ten to some power. Then

$$10^{15} \times 10^? = 10^{30}.$$

Since "exponents add," the question mark must be 15. Hence, the computer takes (more than) 10^{15} seconds.

Most of us don't have much of a feel for how long 10^{15} seconds is. It turns out to be a fairly long time, because there simply are not that many seconds in a year. The calculation that yields the number of seconds in a year can be arrived at by using the commonsense guidance provided by the units as we did before:

$$\#\text{sec./year} = \#\text{sec./min.} \times \#\text{min./hour} \times \#\text{hours/day} \times \#\text{days/year}.$$

Since we are doing a back-of-the-envelope calculation, we shall assume there are 100 seconds in a minute, 100 minutes in an hour, etc. This overestimation gives us:

$$
\begin{aligned}
\#\text{sec./year} &= \#\text{sec./min.} \times \#\text{min./hour} \times \#\text{hours/day} \times \text{days/year} \\
&< \quad 100 \quad \times \quad 100 \quad \times \quad 100 \quad \times \quad 1{,}000 \\
&= \quad 10^2 \quad \times \quad 10^2 \quad \times \quad 10^2 \quad \times \quad 10^3 \\
&= \quad 10^9.
\end{aligned}
$$

Hence, there are fewer than 10^9 seconds in a year.

Now, given that we know

1. The computer takes at least 10^{15} seconds to complete the task, and

2. There are fewer than 10^9 seconds in a year,

we can ask how many years it will take the computer to complete the task. Again, the units show the way:

$$[\# \text{ seconds/year}] \times [\# \text{ years}] = [\# \text{ seconds}].$$

Thus,

$$[10^9] \times [\# \text{ years}] = [10^{15}].$$

So, if the number of years is a power of 10, we have,

$$10^9 \times 10^? = 10^{15}.$$

Again, exponents add, so the question mark must be 6. We can now conclude that (more than) 10^6 years are needed. But $10^6 = 1,000,000$, and so the answer to our question of how long it would take you to do the game-tree analysis (by computer) is:

You couldn't do it in a million years.

■ 6.5 STATEMENT OF O'NEILL'S THEOREM

The previous section suggests that, at least at the time of this writing, the naïve pruning of the game tree to find the optimal opening bid in the (ideal) dollar auction is simply not viable for values of the bankroll b even as small as 120. Nevertheless, the game-tree analysis presented should suffice to convince the reader that such pruning could, in theory, be done, and, hence, there definitely exists an optimal opening bid. But, how do we find it if not by actually pruning the tree? The answer is given by the following elegant theorem of Barry O'Neill (1986):

THEOREM. (O'Neill). For stakes $s > 1$, equal bankrolls $b > 0$, and the conservative convention, the optimal opening bid in the dollar auction can be calculated as follows: One repeatedly subtracts $s - 1$ from b until one more subtraction would yield zero or a negative number. The result of the last subtraction is the optimal opening bid. Rationality implies that if Player 1 opens with this bid, then Player 2 will pass.

We shall illustrate the use of O'Neill's theorem for the case where units are dimes, stakes one dollar ($s = 10$), and bankroll twelve dollars ($b = 120$). Notice that $s - 1 = 9$. Thus, according to the theorem, we simply subtract 9 from 120 as many times as we can without hitting or going below zero. This final result will be our opening bid. The sequence of calculations is as follows:

$$120 - 9 = 111$$
$$111 - 9 = 102$$
$$102 - 9 = 93$$
$$93 - 9 = 84$$
$$84 - 9 = 75$$
$$75 - 9 = 66$$
$$66 - 9 = 57$$
$$57 - 9 = 48$$
$$48 - 9 = 39$$
$$39 - 9 = 30$$
$$30 - 9 = 21$$
$$21 - 9 = 12$$
$$12 - 9 = 3$$

We stop here, since an additional subtraction of 9 would yield -6, which is less than zero. Thus our conclusion, from the theorem, is that the optimal opening bid is three units (i.e., thirty cents). As mentioned earlier, the reader should find this extremely nonobvious.

As one more quick example, if $b = 75$ and $s = 12$, then $s - 1 = 11$, and so the sequence of subtractions would yield:

$$75, 64, 53, 42, 31, 20, 9$$

Thus, the correct opening bid in this case is 9.

The proof of O'Neill's theorem (using a proof technique known as mathematical induction) is given in Chapter 12. One of the exercises at the end of that chapter also asks the reader to use mathematical induction to prove the assertion we made in this section about the number of branches in a binary tree. (We mention this for the sake of the instructor or reader who prefers treatments to be as self-contained as possible.)

We should also point out that in proving O'Neill's theorem in Chapter 12, we are led to consider a stronger version of the theorem that answers the following question: Suppose s = one dollar, b = twelve dollars, units are dimes, and you open with a bid of thirty cents assuming rationality of your opponent. Suppose your opponent (irrationally) says forty cents. What do you do? The answer, of course, depends on whether or not you can count on your opponent to be rational for the rest of the auction. If so, you should bid $1.20. (This should be far from obvious.) If not, you're on your own.

O'Neill (1986) makes a distinction between what he calls the "real dollar auction" and the "ideal dollar auction." The former refers to what takes place among unsuspecting participants in such an auction, while the latter refers to the result of sequential rationality via the kind of analysis treated in this chapter. The distinction is an important one. In fact, one might well argue that the former is of significance to political science, but mathematically uninteresting, and the latter is just the opposite. While we tend to agree with this, we do, however, feel that the dollar auction serves as a striking introduction to the idea of rational analysis based on looking ahead several moves in game-theoretical situations, and that this *is* of importance in political science.

Nevertheless, the question of whether either the real dollar auction or the ideal dollar auction is a "good" model of escalation is one that certainly warrents consideration. In fact, one thing to be considered in judging the value of a model is the quality and quantity of the questions it raises. For the ideal dollar auction and the real dollar auction, O'Neill (1986) has given us exactly such a set of questions. He presents a dozen aspects of escalation (limited ability to look ahead, availability of third-party intervention, etc.) and discusses the models versus real-world escalation in these contexts. We won't, however, reproduce these questions here; there simply is no substitute for going to the literature, and so we refer the reader to O'Neill's article for more on this.

■ 6.6 VICKREY AUCTIONS

No one has ever suggested that the dollar auction is a serious mechanism for selling any kind of goods or service. Indeed, we have discussed it only as a model of escalatory behavior and to illustrate some ideas of fundamental importance in game theory (strategy and rationality) and mathematics in general (game-tree analyses). On the other hand, real-world auctions are not only extremely interesting from a mathematical point of view, but they serve as a mechanism for the transfer of millions of dollars every day. For this reason, we will briefly discuss real-world auctions and a very pretty theorem due to William Vickrey.

One usually distinguishes between two fundamental auction situations. The first type is referred to as a "private-value auction" and is typified by the scenario in which we find ourselves standing in a tent and bidding on a lamp that we have no intention of reselling, but simply want for our own house. What the lamp is worth to us is independent of what everyone else thinks the lamp is worth to them—it is our own private value. The second type of auction situation is called a "common-value auction." This is typified by the bidding of several companies for oil rights. The oil under the particular piece of ground being auctioned off has a definite value and this value is essentially the same for all the bidders. The problem is that none of the bidders knows exactly what this value is because he or she doesn't know how much oil is there.

This dichotomy oversimplifies real life, since auctions tend to be some combination of private value and common value. It turns out that the private-value situation is the easier to analyze, while the common-value situation is more important in terms of the amount of money involved.

In addition to having two types of auctions (private value and common value), there are two kinds of auctions of each type: oral auctions and sealed-bid auctions. Our primary interest here is with sealed-bid auctions. Although one probably tends to think of sealed-bid auctions in the context of construction projects such as a new building on an academic campus or an addition to a private residence, we shall, for simplicity, still focus on the simpler situation where we have several people bidding for a lamp. The difference is in whether the auctioneer is the buyer or seller and, thus, whether the lowest bid wins or the highest bid wins. By sticking with the lamp example, we remain in the more familiar territory where the highest bidder wins.

The first kind of sealed-bid auction mechanism we consider is one in which we ignore the possibilities of ties and the lamp simply goes to the highest bidder who then pays the auctioneer whatever he or she bid. We shall call this a *first-price auction* (for reasons that will become clear in a moment). This is probably the only kind of sealed-bid auction that suggests itself to the reader. Indeed, in spite of the results to be presented, it is by far the most widely used kind of sealed-bid auction.

Intuition suggests that in a first-price sealed-bid auction, a bidder may want to submit a bid that is less than the lamp's actual value to that bidder. In fact, if one formalizes the problem in the standard way (which requires making certain assumptions about such things as the value of the lamp to the other bidders), then it can be proven that (in some sense) the optimal strategy is to submit a bid that is a fraction of your true valuation, and that the appropriate fraction is

$$\frac{n-1}{n}$$

where n is the total number of bidders. Thus, if there are two bidders, one should bid only half of what he or she feels the lamp is really worth; if there are three bidders, two-thirds; four bidders, three-fourths, and so on. Although this is true, it should not be obvious.

In 1961, the economist William Vickrey published one of the truly seminal papers in the theory of auctions (Vickrey, 1961). One of the themes in his paper began with the observation that a first-price sealed-bid auction is "strategically equivalent" to a descending oral auction (also called a "Dutch auction") in which the auctioneer begins at a very high price and gradually lowers the price until someone calls "mine." The main point underlying this observation is that in both auction situations, the bidder must choose how high to bid without knowing the other bidders' decisions. [More on this can be found in the very nice survey papers of Preston McAfree and John McMillan (1987) and Paul Milgrom (1989). We should also point out that small laboratory experiments, according to Milgrom, have suggested that bidders actually tend to pay a lower price in Dutch auctions than in the theoretically equivalent first-price sealed-bid auctions.]

Given that oral descending auctions (Dutch auctions) can be replaced by first-price sealed-bid auctions, Vickery then asked the following: Can one find a sealed-bid procedure that yields the same outcome as an oral ascending auction (English auction)? The answer to this requires a moment's thought as to what price the winner actually pays for the lamp in question. The point is that the winning bidder in an English auction is typically not paying the price that represents the true value of the lamp to this bidder. He or she is paying (slightly more than) the value of the lamp to the *second-highest* bidder. Thus, Vickery introduced the notion of a *second-price sealed-bid auction* (also known today as a *Vickrey auction*): the highest bidder wins the auction, but he or she only pays the second-highest bid.

One question that should occur to the reader is the following: Why would anyone trying to sell something at auction ever contemplate using a second-price sealed-bid auction instead of a first-price sealed-bid auction? Wouldn't the former choice simply mean selling the lamp at a lower price than the latter choice? The answer to this—under some fairly reasonable assumptions—is quite surprising: the expected price paid to the auctioneer is the same in a first-price sealed-bid auction and a second-price sealed-bid auction. This is known as the revenue equivalence theorem, and also goes back to Vickrey's 1961 paper.

So how can this be? The answer lies in the intuition that says we can afford to be more aggressive in the bidding if we know that the price we will pay—if we win—is definitely less than what we bid. How much more aggressive? The answer is given in the following remarkable theorem of Vickrey. It asserts that honesty is, indeed, the best policy in a Vickrey auction, in the sense that one bidding strategy (in this case, honesty) is said to *weakly dominate* another if it is at least as good as the other in every scenario, and strictly better than the other in at least one scenario.

THEOREM. *In a second-price sealed-bid auction (a Vickrey auction), the strategy of bidding one's true valuation for the object being sold weakly dominates every other bidding strategy.*

PROOF. Let v denote the actual value to us of the object being auctioned, and let x denote the highest bid among those made by our competitors. We first show that a bid of v yields an outcome at least as good as does any other bid we might make. (Although not necessary for the proof, we show this is true even if we knew the value of x before making this other bid.) We consider three cases:

Case 1: x is Greater than v

Any bid we might make that is less than x yields the same outcome for us as does a bid of v: a loss of the auction. Any bid we might make that is greater than x yields an outcome that is strictly worse for us than does a bid of v: winning the object but paying more than we think it is worth as opposed to losing the auction. (For a bid of x itself, see Exercise 26.)

Case 2: x is Less than v

Any bid we might make that is less than x yields an outcome that is strictly worse for us than does a bid of v: losing the auction as opposed to winning the auction and paying less than we think the object is worth. Any bid we might make that is greater than x yields the same outcome for us as does a bid of v: winning the object and paying x, which is less than we think it is worth. (For a bid of x itself, see Exercise 26.)

Case 3: x = v

See Exercise 27.

This shows that a bid of v is at least as good as any other bid. We must now show that for any other potential bid there is a scenario where a bid of v yields us a strictly better outcome than does this other bid.

Consider first a potential bid *lo* that is lower than v. If *lo* is less than v − 1 and x = v − 1, then a bid of *lo* yields a loss, whereas a bid of v results in winning the object and paying v − 1. This latter outcome is strictly better than the former. (For the case where *lo* is v − 1, see Exercise 28.)

Consider now a potential bid *hi* that is higher than v. If *hi* is greater than v + 1 and x = v + 1, then a bid of *hi* yields a win but at a higher price than we think the object is worth, whereas a bid of v yields a loss. Again, the latter outcome is strictly better than the former, as desired. (For the case where *hi* is v + 1, see Exercise 29.)

This completes the proof.

In spite of this elegant result (and in spite of the revenue equivalence theorem), Vickrey auctions are almost never used in the real world.

....

■ 6.7 CONCLUSIONS

In this chapter, we presented two theorems. The first, Barry O'Neill's, tells us that the optimal opening bid in the dollar auction (with dimes as units and twelve dollars as bankroll) is thirty cents. The second, William Vickrey's, tells us that the optimal opening bid in a second-price sealed-bid auction is the true value that we attach to the object being sold. The point we want to make here involves a comparison of the sense in which the prescribed behaviors are "optimal."

Suppose, for example, that on the first day of some future semester you find yourself confronted by an instructor who thumbtacks a dollar bill to the strip over the blackboard and announces that he or she will auction off the dollar bill—dime increments and no bid to exceed twelve dollars. Would you open for thirty cents? Suppose that at some future date, you find yourself half-standing in a puddle at the edge of a tent where an old man with a Maine accent announces he is going to hold a second-price sealed-bid auction for a lamp that strikes your fancy. Do you make an honest bid of what the lamp is really worth to you?

Our answers: No and yes.

Why do we come down on different sides of the two questions? Primarily, it is because of the different nature of the assumptions in the two theorems. O'Neill's theorem says we should bid thirty cents if we're rational and everyone else is rational, etc. Moreover, for people to act rationally, they have to be able to do the analysis required to know what their rational strategy is. In the dollar auction, this means pruning the tree (forget it) or knowing O'Neill's theorem. Even if everyone in the class knew O'Neill's theorem, we'd still stay out of the auction all together (or bid ninety-five cents, although we've seen students do this and get burned by peers).

But Vickrey's theorem gives us a weakly dominant strategy. We can't go wrong with it no matter how irrational the other bidders happen to be. Moreover, we're using the phrase *can't go wrong* in a very strong way: if we are honest, when all is said and done, if we were given the chance to change our bid, we wouldn't do it.

Thus, O'Neill's theorem requires that the other players are rational and capable of calculating (in some way) the consequences of rational play on everyone's part. Vickrey's theorem requires no such assumption about one's opponents' rationality or the motives behind their actions.

EXERCISES

1. Do the game-tree analysis for the standard dollar auction with the conservative convention for the case where $s - 3$ and $b = 4$. (The tree will have fifteen terminal nodes, and the analysis will show that the optimal opening bid is *not* 1 unit as was the case for $s = 3$ and $b = 3$.)

2. Do the game-tree analysis for the standard dollar auction with the conservative convention for the case where $s = 3$ and $b = 5$. (The tree for this one is fairly large. It can be done neatly on a standard-size piece of paper, but it requires some care.)

   Variants of the dollar auction for a fixed value of the stakes s and bankroll b can be obtained by varying any of the following three parameters:

   A. The convention used when a place on the game tree is reached for which two different bids are equivalent, in the sense of yielding the same outcome for the player whose turn it is to bid. For example:

      (i)   The conservative convention. This is what we used in **Section 6.3**.

      (ii) The "punishing convention": One chooses between two or more equivalent bids by selecting the one leading to the result least preferred by his or her opponent. If two or more equivalent bids lead to the same outcome for both the bidder and his or her opponent, then the conservative convention is invoked.

   B. The penalty imposed on the second-highest bidder. For example:

      (i)   The second-highest bidder pays what he or she bid. This is what we used in **Section 6.3**.

      (ii) The second-highest bidder pays half of what he or she bid.

   C. The payoff awarded to the highest bidder. For example:

      (i)   The highest bidder receives the stakes (and pays what he or she bid). This is what we used in **Section 6.3**.

      (ii) The highest bidder receives the stakes minus the difference between his or her bid and the second-highest bid (and pays what he or she bid).

These variants are referred to by letter and number in the following exercises. For example, the variant of the dollar auction obtained by using A(i), B(ii), and C(i) is referred to as the "Ai-Bii-Ci dollar auction."

 3. Do the game-tree analysis of the Aii-Bi-Ci dollar auction for the case where (a) $s = 3$ and $b = 3$, and (b) $s = 3$ and $b = 4$.

 4. Do the game-tree analysis of the Ai-Bii-Ci dollar auction for the case where (a) $s = 3$ and $b = 3$, and (b) $s = 3$ and $b = 4$.

 5. Do the game-tree analysis of the Ai-Bi-Cii dollar auction for the case where (a) $s = 3$ and $b = 3$, and (b) $s = 3$ and $b = 4$.

 6. Do the game-tree analysis of the Aii-Bii-Ci dollar auction for the case where (a) $s = 3$ and $b = 3$, and (b) $s = 3$ and $b = 4$.

 7. Do the game-tree analysis of the Aii-Bi-Cii dollar auction for the case where (a) $s = 3$ and $b = 3$, and (b) $s = 3$ and $b = 4$.

 8. Do the game-tree analysis of the Ai-Bii-Cii dollar auction for the case where (a) $s = 3$ and $b = 3$, and (b) $s = 3$ and $b = 4$.

 9. Do the game-tree analysis of the Aii-Bii-Cii dollar auction for the case where (a) $s = 3$ and $b = 3$, and (b) $s = 3$ and $b = 4$.

10. (a) Show that in the dollar auction, it is never rational to increase your previous bid by more than s units. (Do not use O'Neill's theorem.)

    (b) Show that in the dollar auction with the conservative convention, it is never rational to increase your bid by more than $s - 1$ units. (Do not use O'Neill's theorem.)

11. Consider the game played as follows: On move 1, Player 1 chooses one of the three letters: P, Q, R. On move 2, Player 2 chooses one of the two letters: S, T. Finally, on move 3, Player 1 chooses one of the three letters: U, V, W. Player 2 wins if the game results in any of the following sequences: PSW, QSV, QSW, QTU, QTV, RSU, RSV, RSW. Player 1 wins otherwise.

    (a) Draw a tree that shows all possible plays of the game. (The first level will have three nodes, labeled P, Q, and R.)

    (b) Label each terminal node with a "1" or a "2" depending upon whether the corresponding play of the game is a win for Player 1 or Player 2.

    (c) Use backward induction to find which player has a winning strategy.

12. Calculate the exact number of seconds in a year.

13. If it costs 10 cents for every text message you send, and you send 10 messages every weekday and 12 messages every weekend day, how much do your text messages cost you over the course of one (non-leap) year? Assume January 1 is a Monday.

14. According to the World Health Organization, someone in the United States dies of a motor vehicle or traffic accident almost every 5 minutes. How many people in the United States die each year in a motor vehicle or traffic accident?

15. Suppose we have a personal computer (PC) that can do at most 100 million things per second. Suppose we want to do a game-tree analysis of the dollar auction where the units are nickels, the stakes one dollar, and the bankroll four dollars. Show that it would take a reasonably long time. Can it be done in your lifetime?

16. Suppose units are pennies, stakes one dollar, and bankroll four dollars. Do a back-of-the-envelope calculation to show that it would take a very long time to do the game-tree analysis even with a computer that looked at one trillion (=1,000 billion) nodes per second. What does O'Neill's theorem say is the optimal opening bid in this situation?

17. According to O'Neill's theorem, what should the opening bid be in rational play of the dollar auction for the case where stakes are two dollars, bankroll twenty dollars, and units quarters?

18. According to O'Neill's theorem, what should the opening bid be in rational play of the dollar auction for the case where stakes are five dollars, bankroll $100, and units quarters?

19. Construct a chart with 13 columns and 13 rows with the columns labeled 2 through 14 representing values for the bankroll b and the rows labeled 2 through 14 representing values for the stakes s. Use O'Neill's theorem to fill in the chart with optional opening bids.

20. Suppose two dollar auctions are going on simultaneously, with all players bidding according to sequential rationality. Suppose the units are nickels and the stakes one dollar in both cases. Within each auction the bankrolls are fixed and equal, but the value of b for the first auction is five cents more than the value of b in the second auction. Will the opening bids necessarily differ by a nickel? If so, explain why. If not, determine what the maximum possible difference in opening bids (still assuming rational play) could be.

21. Read pages 43 to 49 of O'Neill (1986) and write a two-page paper that both summarizes and comments on what he says.

22. According to O'Neil's theorem, in a dollar auction with stakes $26, the optimal opening bid is $15.
    (a) Give at least five possibilities for what the bankroll might be.
    (b) If you know that the bankroll is between $200 and $225, what must the bankroll equal?

23. Suppose that a dollar auction is held in which the only legal bids that either player can make are 3 units, 4 units, and 6 units. Passing is allowed as usual, with player 1 making the first bid (which cannot be a pass). Winner pays whatever he or she bid and receives the stakes of $s = 7$. Loser pays what he or she bid. Do a game-tree analysis to determine how the auction will rationally proceed.

24. Suppose that a computer can do 100 billion things per second. Use a back-of-the-envelope calculation to estimate how many years it would take the computer to do a game-tree analysis for the dollar auction if the bankroll is 120 units. Assume that the number of seconds in a year is less than 10^9.

25. Suppose that, in the standard dollar auction with the conservative convention, the bankroll is 8, the stakes are 5, and player 1 opens with a bid of 4. Suppose that player 2 irrationally responds with a bid of 6. Assume that from now on, everyone will be rational, follow the

conservative convention, etc. Do a game-tree analysis to determine the optimal response for player 1.

26. In our discussion of Vickrey auctions, we ignored the possibility of ties. In most real-world situations this results in little loss of generality since, for example, the probability of two contractors bidding the same amount for a new library is fairly remote. In fact, one could assign each of the bidders a different number between 0 and 9 and demand that their bid (in dollars) end with that integer.

    On the other hand, suppose we want to deal with ties. Let's agree that if two or more bidders are tied for high bid, a random device decides who will receive the object and the bidder receiving the object pays this tying bid. Since one's true valuation of an object is defined to be the price at which he or she is indifferent between receiving the object at that price and not receiving the object at all, bidders who are honest and find themselves tied for highest bid should be indifferent to the choice made by the random device.

    With this setup, assume x is the highest bid made by our competitors and suppose v is the value to us of the object being auctioned. Show that a bid of v is at least as good as a bid of x
    (a) if $x > v$
    (b) if $x < v$.

27. With the same setup as in Exercise 26, assume $x = v$. Show that a bid of v is at least as good as
    (a) any bid less than v
    (b) any bid greater than v.

28. With the same setup as in Exercise 26, prove that, in a Vickrey auction, bidding one's true valuation v can be strictly better than bidding $v - 1$ (i.e., produce one such scenario).

29. With the same setup as in Exercise 26, prove that, in a Vickery auction, bidding one's true valuation v can be strictly better than bidding $v + 1$.

30. In what ways is a Vickrey auction like an auction on Ebay?

CHAPTER 7

More Social Choice

▪ 7.1 INTRODUCTION

Chapter 1 dealt with what might be called "concrete social choice theory" in that specific social choice procedures were introduced and analyzed. The present chapter deals with what might be called "abstract social choice theory." Rather than consider any particular social choice procedures, this chapter establishes some limitations on what kind of "better" procedures can ever be found. These are very striking results. They are not saying that certain kinds of procedures fail to exist in the sense that no one has yet discovered one, they're saying it is absolutely pointless even to look—one will never be found.

It is more convenient in this chapter to work with so-called social welfare functions (which produce a "social preference list" as output) rather than with social choice procedures (which produce a set of alternatives as output) as we did in Chapter 1. To obtain a still finer analysis, we will speak of a social welfare function for a fixed set of alternatives and/or a fixed set of individuals. (Recall that according to our definition in **Section 1.2**, social choice procedures have to be able to accept as input sequences of individual preference lists corresponding to any set of people and any set of alternatives.) Nevertheless, in **Section 7.2** we show that every social choice procedure as defined in Chapter 1 gives rise to a social welfare function for any set of alternatives and any set of people. (The converse is, in a sense, also true, and much easier to see.)

In **Section 7.3** we state and prove a theorem that generalizes May's theorem (asserting that majority rule is the only "reasonable" voting rule with two alternatives).

In **Section 7.4** we state and prove the single most celebrated result in social choice theory: the impossibility theorem of Kenneth Arrow. The proof we give, although heavily based on ideas developed over the past thirty years, provides a slightly better isolation of the role played by the voting paradox of Condorcet and the condition of independence of irrelevant alternatives.

The second-most prominent result in social choice theory is the Gibbard-Satterthwaite theorem, and we give a proof of this in **Section 7.5**.

Finally, in **Section 7.6**, we consider the relation that suggests itself as the most natural candidate for a social welfare function: place x over y in the social preference list if more than half the individuals have x over y in their individual preference lists. This, in general, won't work, but some fairly natural conditions have been found that—when imposed on the sequences of individual preference lists to be accepted as inputs—guarantee that this will work. Two of the most well-known theorems of this kind, those of Sen and Black, are presented here.

■ 7.2 SOCIAL WELFARE FUNCTIONS

Recall that in **Section 1.2** we defined a social choice procedure to be a function that

1. accepts as input a sequence of individual preference lists of some set A (the set of alternatives), and

2. produces as output either a single alternative (the "winner") or a set of alternatives (i.e., ties were allowed), or the symbol "NW" indicating there is no winner.

The appropriate context for the theorems in this chapter is a slight variant of the above that is usually referred to as a social welfare function. The difference is in the nature of the output. That is, while a typical input is the same in both cases, the output from a social welfare function is not just a single alternative (or a set of alternatives), but a "social preference listing" of the alternatives. As such, it seems to provide more information than a social choice procedure, since one is now being told not only the alternative or alternatives that society considers to be the best, but also the collection that society deems to be in "second place," "third place," etc. The question of whether or not to allow ties (in the preference lists of individuals or the social preference list that is the output) again arises. For our purposes, it seems to be most convenient to follow what we did for social choice functions (i.e., allow ties in the outputs, but not in the inputs). Formalizing this we have the following definition:

DEFINITION. A *social welfare function* is a function that

1. accepts as input a sequence of individual preference lists of some set A (the set of alternatives), and

2. produces as output a listing (perhaps with ties) of the set A. This list is called the *social preference list.*

With social choice procedures and social welfare functions formalized as we have, the apparent generality of the former—and the apparent additional information provided by the latter—turn out to be illusory. This is the content of the following proposition:

PROPOSITION. *Every social welfare function (obviously) gives rise to a social choice procedure (for that choice of voters and alternatives). Moreover (and less obviously), every social choice procedure that always produces a winner gives rise to a social welfare function.*

PROOF. If we have a function that produces as output a listing of all the alternatives, then we obviously can take whichever alternative or alternatives are at the top of the list as the social choice. This shows that every social welfare function gives rise to a social choice procedure (for that choice of voters and alternatives). The second part of the proposition is slightly less obvious.

Suppose, then, that we have a social choice procedure. The question is: How are we going to use this procedure to produce a listing of all the alternatives in A? Obtaining the alternatives that will sit at the very top of the list is easy—they are simply the winners obtained when using the individual preference lists as input for the social choice procedure. The real issue is: How do we determine the group of alternatives that is to occupy "second place" in the social preference list? While not the only possibility to suggest itself, the following may be the most satisfactory: Simply delete from each of the individual preference lists those alternatives that we've already chosen to be on top of the social preference list. Now, input these new individual preference lists to the social choice procedure at hand. The new group of "winners" is precisely the collection of alternatives that we will choose to occupy second place on the social preference list. Continuing this, we delete these "second-round winners" and run the social choice procedure again to obtain the alternatives that will occupy third place in the social preference list, and so on until all alternatives have been taken care of. This completes the proof.

Repeatedly applying a procedure as was done in the second part of the above proof is called "iteration." Some practice in iterating procedures like the Hare procedure and the Borda count is provided in the exercises at the end of the chapter.

Before proceeding, there is one more way in which we want to alter the context in which we will work. In the way that we have formalized both social choice procedures and social welfare functions, individual listings of any set A of alternatives must be allowed as inputs. This permits some fairly strange procedures of either type to be defined. For example, we might designate one person to be the dictator if there are only three alternatives, another if there are four alternatives, and specify that the Borda count be used if there are five or more alternatives. The theorems to follow are best stated in a context that allows for the slightly finer analysis provided by the following definition:

DEFINITION. If A is a set (of alternatives) and P is a set (of people), then a *social welfare function* for A and P is exactly like a social welfare function as defined above except that it accepts as inputs only those sequences of individual preference listings of this particular set A that correspond to this particular set P.

■ 7.3 A GENERALIZATION OF MAY'S THEOREM

We begin by introducing two new properties pertaining to social welfare functions and revisiting an earlier one in this context. Recall that a *permutation* of a collection is simply a rearranging of the collection.

PROPERTY 1. A social welfare function is said to be *anonymous* if it is invariant under permutations of the people. In the general situation, this means that interchanging, say, the fourth list and the seventh list in a sequence of preference lists will not affect what the social preference list looks like. In the context of the two alternatives 0 and 1, this means that whether or not 0 wins depends only upon how many people vote for 0.

PROPERTY 2. A social welfare function is said to be *neutral* if it is invariant under permutations of the alternatives. In the present context, this means, for example, that if:

input: (01101)   yields   output : 0

then

input: (10010)   must yield   output : 1.

That is, if all the ones are changed to zeros and the zeros changed to ones in the input, then the same interchange should take place in the outcome.

PROPERTY 3. In the two alternative context, a social welfare function satisfies *monotonicity* provided that the following holds: If the outcome is 0 and one or more votes are changed from 1 to 0, then the outcome is still 0. We also require that the analogous thing hold for outcome 1.

With these properties (illustrated in the exercises) at hand, we're ready to generalize May's theorem, which says that if the number of people is odd, then majority vote is the only procedure for two alternatives that is anonymous, neutral, monotone, and always produces exactly one of the two alternatives as the social choice. With ties allowed, however, there are others that satisfy anonymity, neutrality, and monotonicity. For example, consider the procedure that simply declares the social choice to be a tie between the two alternatives regardless of how the people vote. More generally, the following collection of procedures provides a spectrum of possibilities between majority vote and the "always tied" procedure just described.

QUOTA SYSTEMS

Suppose we have n people and two alternatives. Fix a number q (for quota) that satisfies

$$\frac{n}{2} < q \le n + 1.$$

Consider the procedure wherein the outcome is a tie unless one of the alternatives has at least q votes. If one of the alternatives has q or more votes, then it alone is the social choice. Such a procedure (i.e., such a social welfare function for a set of two alternatives) will be called a *quota system*.

If n is odd and $q = (n+1)/2$, then the corresponding quota system is just majority vote. At the other extreme, if $q = n + 1$ and there are only n people, then the corresponding quota system is the one described above where the outcome is always a tie (that is, the quota of $n + 1$ can *never* be achieved when there are only n people). It is easy to see that quota systems all satisfy anonymity, neutrality, and monotonicity (see Exercise 4). Much more striking is the following theorem:

THEOREM. *Suppose we have a social welfare function for two alternatives that is anonymous, neutral, and monotone. Then that procedure is a quota system.*

PROOF. Assume we have an arbitrary social welfare function at hand that satisfies the three stated properties. Recall that n denotes the number of people and that we are considering the particular two alternatives

0 and 1. We want to show that there exists a number q satisfying the following two conditions:

1. The alternative 0 alone Is the social choice precisely when q or more people vote for 0.

2. $\frac{n}{2} < q \leq n+1$.

Since we are assuming that the procedure under consideration is invariant under permutations of the people, we know that the outcome depends only on the number of people who vote for, say, 0. Thus, it makes sense to consider the set $\mathcal{G}$ of all numbers k such that 0 alone is the social choice when exactly k people vote for 0. If $\mathcal{G}$ is empty, then 0 alone never wins (and thus 1 alone never wins by neutrality), and so the outcome is always a tie. This is precisely the case $q = n + 1$ described above and we're done in this case. If $\mathcal{G}$ is not empty, then we will let q be the smallest number in the set $\mathcal{G}$.

Notice that monotonicity immediately gives us condition 1 above. To see that condition 2 also holds, notice first that if k is in $\mathcal{G}$ then $n - k$ is definitely not in $\mathcal{G}$. That is, if $n - k$ were also in $\mathcal{G}$, then, by neutrality, we'd get 1 alone as the social choice when exactly $n - k$ people voted for 1. But there are n people all together and so we have exactly k of them voting for 0 precisely when exactly $n - k$ of them vote for 1.

Given this observation about k and $n - k$, we can invoke monotonicity to conclude that if k is in G, then $n - k$ is not as large as k. Thus, $n - k < k$ and so, adding k to both sides, $n < 2k$. Hence, $n/2 < k$ for any number k that is in $\mathcal{G}$. But q is in G, and so we get $n/2 < q$ exactly as desired. This completes the proof.

An immediate consequence of the above is May's theorem from **Section 1.2**.

··

■ 7.4 ARROW'S IMPOSSIBILITY THEOREM

In 1950, Kenneth Arrow published a paper in the *Journal of Political Economy* entitled "A Difficulty in the Concept of Social Welfare." The celebrated result contained in this paper has since become known as Arrow's impossibility theorem. According to Paul Samuelson—himself a Nobel Laureate—the discovery of the impossibility theorem was one of the main reasons that Kenneth Arrow was awarded the Nobel Prize in Economics.

Our goal in this section is to state and prove Arrow's impossibility theorem in the context of social welfare functions. (Recall that this means that a typical input will be a sequence of individual preference lists without ties, and a typical output will be a social preference list, perhaps with ties.) Essentially no new ideas are needed for the extension to the case where ties are allowed in the individual preference lists (see Exercise 11).

Our presentation of Arrow's theorem is divided into three subsections:

1. The statement of Arrow's theorem (including some motivation intended to show that the question being asked is a natural one and that the answer being provided is an extremely surprising and unsettling one).

2. The setup for the proof (where we identify the key concept of "decisiveness" and outline the strategy for the argument to follow).

3. The five lemmas (and the very short proof of each) needed to complete the proof of Arrow's theorem.

The Statement of Arrow's Theorem

Assume for the moment that we have a fixed set A of three or more alternatives and a fixed finite set P of people. Suppose our goal is to find a social welfare function for A and P that is "reasonable" in the sense of reflecting the will of the people. The question is: Exactly what do we mean by "reasonable"? The difficulty in defining such a term lies not so much with generating a list of properties that everyone will agree should be involved, but in trying to overcome the potentially infinite sequence of objections that all begin: "Sure, all those properties should be satisfied by any procedure that we are going to call 'reasonable,' but what about the following additional properties that you haven't mentioned?"

Thus, as a starting point, we might offer a definition of "weakly reasonable" with the understanding that a definition of "reasonable" might later be phrased as "weakly reasonable plus some other conditions." In this spirit, we offer the following definition:

INFORMAL DEFINITION. A social welfare function (for A and P) is called *weakly reasonable* if it satisfies the following three conditions:

1. Pareto: If the input consists of a sequence of identical lists, then this single list should also be the social preference list produced as output.

2. Independence of irrelevant alternatives (IIA): Suppose we have our fixed set A of alternatives and our fixed set P of people, but two different sequences of individual preference lists. Suppose also that exactly the same people have alternative x over alternative y in their list in the first sequence as in their list in the second sequence. (Hence, exactly the same people have alternative y over alternative x in their list.) Then—and this is the conclusion guaranteed by IIA—we either get x over y in both social preference lists, or we get y over x in both social preference lists, or we get x and y tied in both social preference lists. (In other words, the positioning of alternatives other than x and y in the individual preference lists is irrelevant to the question of whether x is socially preferred to y or y is socially preferred to x.)

3. Monotonicity: If we get x over y in the social preference list, and someone who had y over x in his individual preference list interchanges the position of x and y in his list, then we still should get x over y in the social preference list.

With this informal definition at hand, the obvious question now becomes the following:

QUESTION. Are there any weakly reasonable social welfare functions for A and P?

ANSWER. Yes—appoint a dictator.

In the context of social welfare functions, the procedure corresponding to the intuitive idea of a dictator involves not just taking the alternative on top of the dictator's list as the social choice, but rather taking the dictator's entire individual preference listing of A and declaring it to be the social preference list.

Dictatorships are particularly unappealing in the context of trying to reflect the "will of the people." At the moment, however, this is not too unsettling. It would simply seem to mean that when we pass from our definition of weakly reasonable to reasonable, some additional properties will need to be added to rule dictatorships out. Thus, we could simply dismiss the above question, and ask instead the following:

QUESTION. Are there any others?

At this point, however, things become extremely unsettling and our whole proposed program of arriving at "reasonable" social welfare functions crumbles in the wake of the simple (to state) answer provided by Arrow.

ANSWER. No.

THEOREM. *(Arrow, 1950). If A has at least three elements and the set P of individuals is finite, then the only social welfare function for A and P satisfying the Pareto condition, independence of irrelevant alternatives, and monotonicity is a dictatorship.*

REMARKS.

1. It is worth again emphasizing that the theorem is saying that weakly reasonable social welfare functions for A and P other than a dictatorship simply do not exist—not that they haven't yet been found, but that they never will be found.

2. As remarked earlier, the proof we present suffices, with only small modifications, to handle social welfare functions for which ties are allowed in the individual preference lists. On the other hand, in this context the statement of the theorem changes slightly (see Exercise 11 at the end of the chapter).

3. The reference to monotonicity is completely unnecessary. It is included simply because it makes the proof conceptually easier. The additional lemma needed to remove monotonicity from the statement of Arrow's theorem can be found in Exercise 10 at the end of the chapter.

4. Arrow's theorem is referred to as an impossibility result because it can be restated in the following way (where we have stated the stronger version that is indicated by the remarks above). Non-dictatorship, as a property, simply says there is no individual whose placement of x over y in his individual preference list guarantees that x will occur over y in the social preference list.

THEOREM. (*Restatement of Arrow's Theorem*). *If A has at least three elements and the set P of individuals is finite, then it is impossible to find a social welfare function for A satisfying the Pareto condition, independence of irrelevant alternatives, and non-dictatorship.*

The Setup for the Proof

Suppose we have a fixed set A of three or more alternatives, a finite set P of individuals, and a social welfare function for A and P that satisfies Pareto, independence of irrelevant alternatives, and monotonicity. We will produce a dictator. In fact, one way to think about the structure of the proof is to imagine we have before us a "black box" into which we can feed any sequence of individual preference lists, and which will then output a single (preference) list. We don't yet know what goes on inside the black box, but we do know that Pareto, IIA, and monotonicity are always satisfied. Our goal is to show that the "mystery procedure" is, in fact, a dictatorship.

There are four key elements in the setup for the proof: two definitions, a crucial observation, and an easily stated goal that will immediately yield the dictator we want to produce. Keep in mind that the definitions, etc. to follow apply to the particular social welfare function for A and P that we have at hand.

DEFINITION. Suppose X is a set of people, and suppose x and y are two distinct alternatives. Then we'll say that "X can force x over y," or equivalently, "X can force y under x" to mean that we get x over y in the social preference list whenever everyone in X places x over y in their individual preference lists.

Notice that we are not saying that the voters in X will *necessarily* choose to rank x over y in any particular election. All we are saying is that *if* they choose to do so, then we'll definitely get x over y in the social preference list.

CRUCIAL OBSERVATION. Because the fixed social welfare function for A and P that we are considering satisfies independence of irrelevant alternatives and monotonicity, it is much easier to show that a given set X forces some alternative x over some other alternative y than it first appears. That is, in order to show that X forces x over y it suffices to produce a *single sequence* of individual preference lists for which the following all hold:

1. Everyone in X has x over y in their lists.

2. Everyone not in X has y over x in their lists.

3. The resulting social preference list has x over y.

To see why this suffices, note first that independence of irrelevant alternatives says that whether or not we get x over y in the social preference list does not depend in any way on the placement of other alternatives in the individual preference lists. Hence, in showing that X forces x over y, it suffices to consider a single sequence of individual preference lists with all other alternatives strategically placed where we want in order for our argument to go through. Condition 2 above merely reflects the fact that monotonicity says it suffices to consider the "worst case scenario" in which everyone not in X tries to prevent having x over y in the social preference list by placing y over x in their individual preference lists.

We should point out that the standard terminology for "X can force x over y" is "X is decisive for x against y."

DEFINITION. A set X will be called a *dictating* set if X can force x over y whenever x and y are two distinct alternatives in A.

The reader can check his or her familiarity with the above two definitions and crucial observation by verifying the following (see Exercise 7):

1. If X is the set of *all* individuals, then X is a dictating set.

2. If p is one of the individuals and X is the set consisting of p alone, then X is a dictating set if and only if p is a dictator.

Small dictating sets are intuitively harder to come by than large ones. (In fact, in the presence of IIA, this is essentially what monotonicity says.) Pareto and IIA start us off by guaranteeing that the best possible candidate for a dictating set—namely $X = P$—indeed is a dictating set. The conclusion we want to reach is at the other extreme. That is, we want to find the smallest possible kind of dictating set, one with only one element. The strategy for passing from the very large dictating set P where we are starting to the very small dictating set $\{p\}$ where we want to end up involves the following:

GOAL. Show that if X is a dictating set, and if we split X into two sets Y and Z (so that everyone in X is in exactly one of the two sets), then either Y is a dictating set or Z is a dictating set.

In the next part of this section, we'll prove five lemmas, the last of which is precisely what we have stated as our goal. The proof that the goal yields the desired dictator is left to the reader (see Exercise 8 at the end of the chapter)

Before getting to the five lemmas, however, let's prove one proposition that we could actually do without, but at the expense of a slight complication in the five lemmas that follow. This proposition provides a nice opportunity to see how independence of irrelevant alternatives and Pareto are used together.

PROPOSITION. If A has at least three elements, then any social welfare function for A that satisfies both independence of irrelevant alternatives and the Pareto condition will never produce ties in the output.

PROOF. Assume, for contradiction, that we have a social welfare function for A that satisfies both independence of irrelevant alternatives and the Pareto condition, and that some sequence of individual preference lists results in a social preference list in which the alternatives a and b are tied, even though we are not allowing ties in any of the individual preference lists. Because of independence of irrelevant alternatives, we know that a and b will remain tied as long as we don't change any individual preference list in a way that reverses that voter's ranking of a and b.

Let c be any alternative that is distinct from a and b. Let X be the set of voters who have a over b in their individual preference lists, and let Y be the rest of the voters (who therefore have b over a in their lists). Thus, in notation that should be self-explanatory, we have:

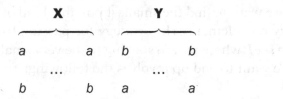

yields

$$ab \quad \text{(tied)}.$$

Suppose we now insert c between a and b in the lists of the voters in X, and we insert c above a and b in the lists of the voters in Y. Then we will still get a and b tied in the social preference list (by independence of irrelevant alternatives), and we will get c over b by Pareto, since c is over b in every individual preference list. Thus, we have:

$$
\begin{array}{cccc}
& \overbrace{}^{X} & & \overbrace{}^{Y} \\
a & a & c & c \\
c \quad \cdots & c & b \quad \cdots & b \\
b & b & a & a
\end{array}
$$

yields

$$
\begin{array}{c}
c \\
ab.
\end{array}
$$

Independence of irrelevant alternatives guarantees us that, as far as *a* versus *c* goes, we can ignore *b*. Thus, we can conclude that if everyone in *X* has *a* over *c* and everyone in *y* has *c* over *a*, then we get *c* over *a* in the social preference list.

To get our desired contradiction, we will go back and insert *c* differently from what we did before. This time, we will insert *c* under *a* and *b* for the voters in *X*, and between *a* and *b* for the voters in *Y*. Using Pareto as before shows that we now get:

X			**Y**		
a		*a*	*b*		*b*
b	...	*b*	*c*	...	*c*
c		*c*	*a*		*a*

yields

$$ab$$
$$c.$$

Once again, independence of irrelevant alternatives guarantees us that, as far as *a* versus *c* goes, we can ignore *b*. Thus, we can now conclude that if everyone in *X* has *a* over *c* and everyone in *Y* has *c* over *a*, then we get *a* over *c* in the social preference list. This is the opposite of what we concluded above, and thus we have the desired contradiction.

The cognoscenti should note that an immediate consequence of this proposition and the voting paradox profile is that there is no social welfare function satisfying Parato, independence of irrelevant alternatives, and invariance under permutations of the alternatives (defined in the natural way).

The Five Lemmas Yielding Arrow's Theorem

The following five lemmas will suffice to complete the proof of Arrow's theorem. Notice that independence of irrelevant alternatives is directly appealed to only in Lemma 1, and this is where the role played by the voting paradox of Condorcet is clearly displayed.

LEMMA 1. *Suppose X forces a over b, and c is an alternative distinct from a and b. Suppose now that X is split into two sets Y and Z (either of which may be the empty set) so that each element of X is in exactly one of the two sets. Then either Y forces a over c or Z forces c over b.*

(Intuition: If X has the power to force a high and b low, then either Y inherits the power to force a high or Z inherits the power to force b low.)

PROOF. Consider what happens when the social welfare function under consideration is applied to the following sequence of individual preference lists:

Everyone in Y	Everyone in Z	Everyone else
a	c	b
b	a	c
c	b	a

(Alternatives other than a, b, and c can be placed arbitrarily in the individual preference lists.) Notice that everyone in both Y and Z (and thus everyone in X) has a over b. Since we are assuming that X forces a over b, this means that we get a over b In the social preference list. In particular, this means that we can't get both b over c and c over a in the social preference list, or else transitivity would give us b over a in the social preference list, contrary to what we just said. Thus, we must have either a over c or c over b in the social preference list. We consider these two cases separately.

Case 1: We Get a Over c In the Social Preference List

In this case, we have produced a single sequence of individual preference lists for which everyone in Y has a over c in their lists, everyone not in Y has c over a in their lists, and the resulting social preference list has a over c. By the crucial observation of the last subsection, this suffices to show that Y forces a over c, and so the lemma is proved in case 1.

Case 2: We Get c Over b In the Social Preference List.

An argument completely analogous to that in case 1 (and left for the reader as Exercise 9 at the end of the chapter) shows that Z forces c over b in this case. This completes the proof of Lemma 1.

LEMMA 2. *Suppose X forces a over b and c is an alternative distinct from a and b. Then X forces a over c and X forces c over b.*

(Intuition: If X can force a over b, equivalently, X can force b under a, then X can force a over anything and X can forces b under anything.)

PROOF. Consider first the special case of Lemma 1 where Y is the whole set X and Z is the empty set. The conclusion is then that either X forces a over c (as desired) or the empty set forces c over b (which is ruled out by the Pareto condition.) Thus X forces a over c. In a completely analogous way, a consideration of the special case of Lemma 1 where Y is the empty set and Z is the whole set X shows that X forces c over b.

LEMMA 3. *If X forces a over b, then X forces b over a.*

(Intuition: The forcing relation is symmetric.)

PROOF. Choose an alternative c distinct from a and b. (This is possible since we are assuming that we have at least three alternatives.) Assume that X forces a over b. Then, by Lemma 2, X forces a over anything. In particular, X forces a over c. But Lemma 2 now also guarantees that X forces c under anything—in particular, X forces c under b. This is the same as saying X forces b over c. Thus, by Lemma 2 one more time, we have that X forces b over anything, and so X forces b over a as desired. Briefly,

$$X \text{ forces } \frac{a}{b} \Rightarrow X \text{ forces } \frac{a}{c} \Rightarrow X \text{ forces } \frac{b}{c} \Rightarrow X \text{ forces } \frac{b}{a}.$$

LEMMA 4. *Suppose there are two alternatives a and b so that X can force a over b. Then X is a dictating set.*

(Intuition: If X has a little local power, then X has complete global power.)

PROOF. Assume X can force a over b, and assume x and y are two arbitrary alternatives. We must show that X can force x over y. Notice that Lemma 3 guarantees that X can also force b over a. Thus, Lemma 2 now lets us conclude that X can force a over or under anything and X can force b over or under anything.

Case 1: a = y

Here, we want to show that X can force x over a. But since we know X can force a under anything, we have that X can force a under x. Equivalently, X can force x over a, as desired.

Case 2: a ≠ y

Since X forces a over b and $a \neq y$, we know that X can force a over y. Equivalently, X can force y under a, and thus X can force y under anything. In particular, X can force y under x. Thus, X can force x over y as desired Briefly,

$$X \text{ forces } \begin{smallmatrix} a \\ b \end{smallmatrix} \Rightarrow X \text{ forces } \begin{smallmatrix} a \\ y \end{smallmatrix} \Rightarrow X \text{ forces } \begin{smallmatrix} x \\ y \end{smallmatrix}.$$

LEMMA 5. Suppose that X is a dictating set and suppose that X is split into two sets Y and Z so that each element of X is in exactly one of the two sets. Then either Y is a dictating set or Z is a dictating set.

PROOF. Choose three distinct alternatives a, b, and c. Since X is a dictating set, we have that X can force a over b. Lemma 1 now guarantees that either Y can force a over c (in which case Y is a dictating set by Lemma 4) or Z can force c over b (in which case Z is a dictating set by Lemma 4 again). This completes the proof of Arrow's impossibility theorem.

■ 7.5 THE GIBBARD-SATTERTHWAITE THEOREM

During presidential elections, we often hear mention of "spoiler candidates," third party candidates who take away votes from the Democratic and Republican nominees. For example, in 2000 and 2004, Ralph Nader supporters were blamed by some Democratic voters for giving the election to George W. Bush. Many Nader supporters were torn between voting for their sincere first choice candidate or voting for their second choice candidate who had a better chance of winning the election. Situations like these bring up the issue of manipulability of voting systems—can a voter obtain a better outcome in an election by voting insincerely? In this section, we discuss and prove the main result regarding the manipulability of voting systems—the Gibbard-Satterthwaite theorem.

In the early 1970s, the theorem was proven independently by philosopher Allan Gibbard and economist Mark Satterthwaite. The theorem is similar to Arrow's theorem in many ways, and in fact, our proof mirrors that of Arrow's theorem at every step. Before we discuss the statement and proof, we will make more precise what we mean by manipulability. Throughout our discussion, we consider only social choice procedures in which ties are not allowed on ballots and in which there is exactly one alternative in the resulting social choice set. Notice that Condorcet's method is not guaranteed to produce a winner, so our treatment of manipulability does not apply (see the exercises, however, for an analysis of Condorcet's method with respect to an alternate definition of manipulability).

DEFINITION. A social choice procedure is *manipulable* if there exists a profile $P = (B_1, \ldots, B_n)$ which we think of as representing the true preference lists of the n voters and another ballot B_i' representing the insincere ballot of a single voter i such that voter i prefers the social choice produced by profile P' (obtained from P by switching the sincere preference list B_i of voter i with the insincere preference list B_i') to the social choice produced by profile P. If a social choice procedure is not manipulable, we say that it is *non-manipulable*.

Despite our intuition about the manipulability of the plurality system, arising from scenarios like the 2000 and 2004 presidential elections described above, plurality is non-manipulable under the definition above when exactly one alternative appears in the social choice set. Let's consider the special case where there are two alternatives, and an odd number of voters. The fact that there are an odd number of voters prevents ties, and guarantees that there will be exactly one social choice for any given profile. Since first place rankings are the only information taken into account and there are only two alternatives, then an insincere ballot is a first place vote for one's second choice alternative. Voting for the opposing alternative will not yield a better result, and so, the social choice procedure is non-manipulable under these conditions. A similar argument (see Exercise 17) shows that plurality is non-manipulable whenever ties are not allowed in the outcome. If we do allow ties, however, or we consider the idea of manipulability by a group rather than a single individual, then it turns out that plurality is manipulable (see Exercises 18 and 19).

Another example of a non-manipulable social choice procedure is a dictatorship. Since only the dictator's preference list matters, an insincere ballot from a voter that is not the dictator has no effect on the outcome. An insincere ballot from the dictator will not yield a better result for the dictator, since the alternative at the top of the dictator's list is the social choice. Recall that a dictatorship also satisfies the Pareto condition and a dictatorship never produces a tie.

QUESTION. Are there other social choice procedures that never yield a tie, satisfy Pareto, and are non-manipulable?

ANSWER. No.

Already we see the parallels with Arrow's theorem. There are no problems for plurality when only two alternatives are being considered (although in this context, we do require that the number of voters be odd), and a dictatorship also satisfies the desired properties (namely, non-manipulability in this context). Unfortunately, the Gibbard-Satterthwaite theorem, like Arrow's theorem, tells us that for three or more alternatives, a dictatorship is the only social choice procedure that satisfies the desirable properties of Pareto and non-manipulability when the social choice procedure under consideration always produces a single alternative as the social choice.

THEOREM (*Gibbard-Satterthwaite, 1973*). *If the set A of alternatives has at least three elements, and the set P of individuals is finite, then the only social choice procedure for A and P which outputs exactly one social choice and satisfies both the Pareto condition and non-manipulability is a dictatorship.*

REMARKS.

1. Just as was the case with Arrow's theorem, the proof we will present of the Gibbard-Satterthwaite theorem suffices with some modifications to treat the situation when ties are allowed in the ballots (see Exercise 20).

2. The full strength of the Pareto condition is not required for the theorem. The result still holds if we replace Pareto with the strictly weaker condition of non-imposition (see Chapter 1, Exercise 21) which is satisfied if every alternative occurs as the unique winner for at least one set of ballots (see Exercise 22).

3. The Gibbard-Satterthwaite theorem is one more example of the impossibility theorems that we have encountered. Non-dictatorship, in this context, means that there is no single individual that can guarantee that alternative *a* is the lone social choice by placing *a* at the top of his or her individual preference list. An alternative version of the theorem then is the following.

THEOREM (Restatement of the Gibbard-Satterthwaite Theorem). *If A has three or more elements, and the set P of individuals is finite, then it is impossible to find a social choice procedure satisfying the Pareto condition, non-manipulability, and non-dictatorship, and which always yields a unique winner.*

The Setup for the Proof

The setup for the proof of the Gibbard-Satterthwaite theorem is remarkably similar to that of Arrow's theorem. Suppose that we have a fixed set A of three or more alternatives, a finite set P of individuals, and a social choice procedure for A and P that satisfies Pareto, non-manipulability, and always outputs exactly one social choice. We will produce a dictator. Just as with Arrow's theorem, we will use the idea of a dictating set. We will show that the set P of all individuals is a dictating set, and we can continue to break P up into smaller and smaller sets until we have a dictating set consisting of a single individual—that individual is the dictator.

The most significant difference in the statements of Arrow's and Gibbard-Satterthwaite's theorem is that Arrow's theorem was stated in the context of social welfare functions, and the Gibbard-Satterthwaite theorem is stated in the context of social choice procedures. Since the social choice procedure produces a single social choice rather than a ranking of alternatives, it no longer makes sense to talk about a set X of voters *forcing* alternative x over alternative y. Instead, we introduce the following notion to take its place.

> **DEFINITION.** Suppose X is a set of people, and x and y are two distinct alternatives. Then we'll say that "X can use x to block y" to mean that y is not the social choice whenver everyone in X places x over y in their individual preferences lists.

Instead of appealing directly to the fact that our given social choice procedure is non-manipulable, we will rely on another property that all non-manipulable social choice procedures satisfy: *down-monotonicity*.

> **DEFINITION.** A social choice procedure is said to satisfy *down-monotonicity* if the following holds for any distinct alternatives x and y:
>   If y is the social choice and x is not the social choice, and someone changes his or her preference list by moving x down one spot (that is, exchanging x's position with that of the alternative immediately below x on his or her preference list), then y should still be the social choice.

It turns out that any social choice procedure that satisfies down-monotonicity must also satisfy monotonicity (see Exercise 12 at the end of the chapter). What is important to us here, however, is that any non-manipulable social choice procedure (that produces exactly one social choice) must satisfy down-monotonicity; it is this property that we will rely on in the proof of the Gibbard-Satterthwaite theorem since it is easier, in practice, to apply than non-manipulability.

> **PROPOSITION.** *Any non-manipulable social choice procedure that produces a single social choice must satisfy the down-monotonicity condition.*

PROOF. Assume that the social choice procedure fails to satisfy down-monotonicity. Then there must exist two profiles P and P' and alternatives x, y, v, and w such that:

(1) Alternative w ($w \neq y$) is the social choice for profile P, and some fixed individual i ranks y immediately above x in his or her preference list;

(2) Alternative v ($v \neq w$) is the social choice for profile P', and P' differs from P only in that individual i has interchanged the positions of x and y on his or her ballot.

We will show that the social choice procedure is manipulable in this situation, a contradiction. It must be the case, therefore, that our social choice procedure does indeed satisfy the down-monotonicity condition. We will consider three cases.

Case 1: In profile P, voter i ranks v above w.

Suppose that voter i's ballot in profile P represents the sincere preferences of voter i, and the ballot in P' is disingenuous. If voter i submits his or her sincere ballot, then w is the social choice, but if voter i submits the disingenuous ballot, then v is the social choice. Since voter i prefers v to w, then voter i is better off submitting the insincere preference list. Thus, in this case, the social choice procedure is manipulable.

Case 2: In profile P', voter i ranks w above v.

Suppose now that voter i's ballot in profile P' represents the sincere preferences of voter i, and the ballot in P is disingenous. If voter i submits his or her sincere ballot, then v is the social choice, but if voter i submits the disingenous ballot, then w is the social choice. Since voter i prefers w to v, then voter i is again better off by submitting the insincere preference list. Again, this shows that the social choice procedure is manipulable.

Case 3: Otherwise.

In this case, we must have that in profile P, voter i ranks w above v, and in profile P', voter i ranks v above w. It must be the case then that

$y = w$ and $x = v$ since x and y are the only alternatives that voter i is switching. But then y is the social choice for profile P, a contradiction, so case 3 is impossible. That completes the proof.

CRUCIAL OBSERVATION We can use the same strategy to show that X can use x to block y that we used to show that X can force x over y. This is not surprising since the two conditions are only different in so far as one makes sense for social choice procedures and one makes sense for social welfare functions. So to show that X can use x to block y, it suffices (see Exercise 14) to produce a single sequence of individual preference lists for which the following all hold:

1.   Everyone in X has x over y in his or her preference list.

2.   Everyone not in X has y over x in his or her preference list.

3.   The resulting social choice is x.

As before, condition 2 reflects our consideration of the worst case scenario; if some of the voters not in X also rank x over y in their preference lists, then by monotonicity, the social choice is still x. Since we are assuming that our social choice procedure is non-manipulable, it is therefore down-monotonic. Since down-monotonicity implies monotonicity, this strategy will always work to show that a set of voters X can use alternative x to block alternative y.

DEFINITION. A set X of voters is called a *dictating* set if X can use x to block y whenever x and y are distinct alternatives in A.

Just as with our previous definition of dictating set, the following properties hold. We leave their verification to the reader.

1.   The set X of all voters is a dictating set.

2.   If p is one of the invidivuals and X is the set consisting of p alone, then X is a dictating set if and only if p is a dictator.

We are now ready to prove five lemmas that will allow us to prove the Gibbard-Satterthwaite theorem. The lemmas are completely analogous to four of the five used to prove Arrow's theorem. The strategy is the same: we know that the set X of all individuals is a dictating set. We will show that when we break X up into two smaller disjoint sets, then one of the two sets must be a dictating set. Repeating this procedure, we eventually obtain a dictating set consisting of a single voter—that voter is the dictator.

The Five Lemmas Yielding the Gibbard - Satterthwaite Theorem

LEMMA 1. *Suppose X can use a to block b, and c is an alternative distinct from a and b. Suppose that X is split into two sets Y and Z (either of which may be the empty set) so that each element of X is in exactly one of the two sets. Then either Y can use a to block c or Z can use c to block b.*

PROOF. We will apply the social choice procedure under consideration to any profile in which everyone in Y ranks a, b, c (in that order) as their top three alternatives, everyone in Z ranks c, a, b as their top three alternatives, and everyone else ranks b, c, a as their top three alternatives.

Everyone in Y	Everyone in Z	Everyone else
a	c	b
b	a	c
c	b	a

By Pareto, the social choice must be one of the alternatives a, b, c (since everyone prefers alternative a (or b or c) to a fourth alternative, say d, then d is not the social choice). We also know that alternative b can not be the social choice since everyone in X ranks a over b and, by assumption, X can use a to block b. There are two possibilities to consider then.

Case 1: Alternative a is the social choice.

In this case, we must have that Y can use a to block c since everyone in Y ranks a over c, everyone not in Y ranks c over a, and a is the social choice.

Case 2: Alternative c is the social choice.

In this case, we must have that Z can use c to block b since everyone in Z ranks c over b, everyone not in Z ranks b over c, and c is the social choice.

Notice that by the crucial observation above, it is sufficient to consider only profiles of the type described above. That completes the proof.

LEMMA 2. *Suppose X can use a to block b, and c is an alternative distinct from a and b. Then X can use a to block c, and X can use c to block b.*

PROOF. We will apply Lemma 1 in the special case that Y is all of X and Z is the empty set. Lemma 1 then implies that either Y can use a to block c or Z can use c to block b. But Z is the empty set, and by Pareto, it is impossible for an empty set of voters to use c to block b. Thus Y can use a to block c as desired. A completely analogous argument to the one above, wherein Y is the empty set and Z is all of X, shows that X can use c to block b as well.

LEMMA 3. *If X can use a to block b, then X can also use b to block a.*

PROOF. Choose an alternative c distinct from both a and b; this is possible since A has at least 3 elements. Since X can use a to block b, then Lemma 2 implies that X can use a to block any alternative. Thus X can use a to block c. Applying Lemma 2 again shows that X can use any alternative to block c; in particular, X can use b to block c. One more application of Lemma 2 shows that X can use b to block anything; thus X can use b to block a, as claimed.

LEMMA 4. *Suppose there are two alternatives a and b for which X can use a to block b. Then X is a dictating set.*

PROOF. Let x and y be any distinct alternatives, and we will show that X can use x to block y. Since x and y are arbitrary, this shows that X is a dictating set.

Case 1: $y \neq a$

Since X can use a to block b, and $y \neq a$, Lemma 2 implies that X can use a to block y. Thus, again by Lemma 2, X can use any alternative to block y; in particular, X can use x to block y.

Case 2: $y = a$

Since X can use a to block b, then Lemma 2 implies that X can use b to block a. By Lemma 2, this means that X can use anything to block a; in particular, X can use x to block a. Since $a = y$, this proves the claim.

LEMMA 5. *Suppose that X is a dictating set and that X is split into two sets Y and Z so that every element of X is contained in exactly one of Y and Z. Then either Y is a dictating set or Z is a dictating set.*

PROOF. We leave the proof as an exercise (see Exercise 15). The argument is analogous to that used to prove Lemma 5 in the proof of Arrow's theorem.

....

■ 7.6 SINGLE PEAKEDNESS—THEOREMS OF BLACK AND SEN

Suppose we have our fixed set A of alternatives and our fixed set P of people. Intuition may well suggest the following as the most natural way to construct a single social preference list given a sequence of individual preference lists:

Place x over y in the social preference list if more than half of the people have x over y in their individual preference lists. (And place x and y tied if exactly half have x over y and half have y over x.)

What's wrong with this?

The answer goes back one more time to the voting paradox of Condorcet. That is if we apply the above procedure to the lists

$$
\begin{array}{ccc}
a & c & b \\
b & a & c \\
c & b & a
\end{array}
$$

we would get a over b in the social preference list as well as b over c and c over a in the same list. This yields a relation among the alternatives, but it's certainly not a list, and our use of the word *over* is misleading at best.

Intuition again suggests that if we have a sequence of individual preference lists where each list is similar (in some sense) to all the others, then the natural way of defining a social welfare function described above may, in fact, work. In this section, we investigate two ways to formalize this notion of "similar to each other." The first is a theorem due to Sen (1966); the second is a classic (although weaker) result due to Black (1958).

The concept occurring in Sen's theorem that captures this idea of a collection of individual preference lists being "similar to each other" is given by the following definition:

DEFINITION. A sequence of individual preference lists will be called *Sen coherent* if for every triple of alternatives there is at least one of the three—call it *x*—such that at least one of the following holds:

1. No one places *x* above both of the other alternatives in the triple.

2. No one places *x* between the other two alternatives in the triple.

3. No one places *x* below both of the other alternatives in the triple.

Example:

We'll go through the procedure for determining if the following sequence of four individual preference lists is Sen coherent.

$$
\begin{array}{cccc}
a & d & a & c \\
b & c & d & b \\
c & b & c & a \\
d & a & b & d
\end{array}
$$

There are four triple of alternatives that we must consider. For each triple, we'll check to see if we can find one of the alternatives in the triple and one of the positions "above the other two," "between the other two," or "below the other two" with the property that the specified alternative does not occur in the specified position. If we can't find such an alternative and such a position (i.e., if they don't exist), then we can conclude that this triple serves as witness to the fact that the sequence of lists is not Sen coherent, and we're done. If we can find such an alternative and position, then we move on to check the next triple. If all triples "check out," we will conclude that the sequence of lists is Sen coherent.

1. First triple: {a, b, c}

    Does anyone place a above both b and c? Yes, person # 1 does.

    Does anyone place a between b and c? No.

    Conclusion: {a,b,c} "checks out." We move on to the next triple.

2. Second triple: {a, b, c}

    Does anyone place a above both b and d? Yes, person # 1 does.

    Does anyone place a between b and d? Yes, person # 4 does.

    Does anyone place a below both b and d? Yes, person # 2 does.

    Does anyone place b above both a and d? Yes, person # 4 does.

    Does anyone place b between a and d? Yes, person # 1 does.

    Does anyone place b below both a and d? Yes, person # 3 does.

    Does anyone place d above both a and b? Yes, person # 2 does.

    Does anyone place d between a and b? Yes, person # 3 does.

    Does anyone place d below both a and b ? Yes, person # 1 does.

    Conclusion: {a, b, d} serves as witness to the fact that the sequence of lists is *not* Sen coherent, since each of the three alternatives occurs in all three positions.

NOTATION. Suppose we have a fixed sequence of individual preference lists and suppose that x and y are two alternatives. Then we will write

$$xPy$$

(read as "x is socially preferred to y") to mean that more than half the people have x over y in their individual preference lists.

The notation above is simply providing a name for the relation that naturally arises when trying to generate a "social preference list." The preference lists involved in the voting paradox yield a situation where we have aPb and bPc and cPa. This violates the condition known as transitivity.

The importance of transitivity as a property is demonstrated in the following proposition. An outline of the proof occurs as Exercise 27 at the end of the chapter.

PROPOSITION. *If the number of people is odd and the relation P is transitive, then it is possible to arrange the alternatives in a vertical list so that for any two alternatives x and y, x is placed higher than y in the list precisely when xPy holds.*

We are now ready to prove the main result of this section. The following proof, which differs from that in Sen (1966), easily generalizes to handle the version of the theorem that allows ties in the individual preference lists. It should also be noted that this proof is not as concise as possible (see Exercise 28 at the end of the chapter), but it is the most direct approach that a naïve attack produces.

THEOREM. *(Sen, 1966). Suppose that the number of individuals is odd and that we have a sequence of individual preference lists that is Sen coherent. Then the relation P is transitive.*

PROOF. Assume that the number of people is odd and that we have a sequence of individual preference lists that is Sen coherent. We want to show that P is transitive. Thus, we assume that we have three arbitrary alternatives x, y, and z, and that xPy and yPz both hold. (That is, more than half the people have x over y in their individual preference lists, and more than half have y over z.) We must show that xPz also holds. Since the sequence of lists at hand is Sen coherent, we know that at least one of the three alternatives (i.e., x, y, or z) fails to occur in at least one of the three positions (i.e., above the other two, between the other two, or below the other two). This yields nine separate cases to be considered. We'll show that in each of the nine, either the case can't occur (because of xPy and yPz) or that we get xPz as desired. This will complete the proof.

For each of the following nine cases we will use the same notation. That is, we'll introduce sets of people denoted by A, B, C, and D (e.g., A might be the set of all people who place x over y in their list). The corresponding lower case letters will denote the number of people in the set. (So a is the number of people in A, b is the number in B, etc.) Typically, then, the total number of people will be given by $a + b + c + d$. Thus, if we have A as above and xPy, then we'll know that $a > (a + b + c + d)/2$.

Case 1: No One has Alternative x Above the Other Two

In this case everyone has x either in the middle of y and z or below y and z. Thus, we can decompose the set of people into four pairwise disjoint sets as follows:

A is the set of people ranking y over x over z.
B is the set of people ranking z over x over y.
C is the set of people ranking y over z over x.
D is the set of people ranking z over y over x.

Notice that only the people in B have x over y. But we are assuming that xPy, and so b is greater than $(a + b + c + d)/2$. Similarly, since yPz we have that $a + c$ is greater than $(a + b + c + d)/2$. But then adding the left-hand sides and the right-hand sides of these two inequalities yields $a + b + c$ is greater than $a + b + c + d$. This would mean that d is negative, which, of course, it's not. Thus, case 1 cannot occur.

Case 2: No One has Alternative x Between the Other Two.

In this case, everyone has x either above y and z or below y and z. Thus, we can decompose the set of people into four pairwise disjoint sets as follows.

A is the set of people ranking x over y over z.
B is the set of people ranking x over z over y.
C is the set of people ranking y over z over x.
D is the set of people ranking z over y over x.

Notice that only the people in A and B have x over y. Thus, since we are assuming that xPy, we know that $a + b > (a + b + c + d)/2$. But everyone in A and B also has x over z. Thus, we get xPz as desired.

Case 3: No One has Alternative x Below the Other Two

In this case, everyone has x either above y and z or between y and z. Thus, we can decompose the set of people into four pairwise disjoint sets as follows.

A is the set of people ranking x over y over z.
B is the set of people ranking x over z over y.
C is the set of people ranking y over x over z.
D is the set of people ranking z over x over y.

Notice that only the people in A and C have y over z. Thus, since we are assuming that yPz, we know that $a + c > (a + b + c + d)/2$. But everyone in A and C also has x over z and so we get xPz as desired.

The Remaining Cases

The remaining six cases are left as an exercise for the reader (see Exercise 29 at the end of the chapter). We'll content ourselves here to simply list what the cases are and to note the conclusions that will arise: cases 5 and 9 are like case 1 in that they also can't occur; cases 4, 6, 7, and 8 yield the conclusion that xPz as desired.

Case 4: No One has Alternative y Above the Other Two
Case 5: No One has Alternative y Between the Other Two
Case 6: No One has Alternative y Below the Other Two
Case 7: No One has Alternative z Above the Other Two
Case 8: No One has Alternative z Between the Other Two
Case 9: No One has Alternative z Below the Other Two
This completes the proof of Sen's theorem.

Sen's theorem arose as a generalization of a classic result first proved in 1948 by Duncan Black. Black's Theorem provides an answer to the same question as does Sen's theorem: Can we find a natural condition pertaining to sequences of individual preference lists that will guarantee the transitivity of the relation P, where xPy means that more then half the people rank x over y? Instead of what we called "Sen coherence," Black's theorem makes use of a coherence property based on the idea of "graphing" an individual preference list. We illustrate this with the following example:

Example:

Suppose the set of alternatives is {a, b, c, d} and consider the fixed ordering $d\ b\ a\ c$ of these alternatives. Consider also the individual preference list in which we have a over b over d over c. Then a graph of this list with respect to this ordering is obtained in the following way:

1. Place the alternatives on a horizontal line arranged from left to right as in the given fixed ordering.

2. Place a dot above each alternative so that the dot above alternative x is placed higher than the dot above alternative y precisely when alternative x occurs above y in the given individual preference list.

3. (Optional) Connect each dot by a line to the ones immediately to the left and right of it.

Notice that both of the following are graphs of the above list with respect to the above given order.

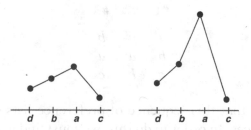

DEFINITION. A *peak* in a graph (arrived at as above) is a dot that is higher than the dot immediately to its left (if there is one to its left) and higher than the dot immediately to its right (if there is one to its right).

Example:

The three possible kinds of peaks are illustrated in the following figure:

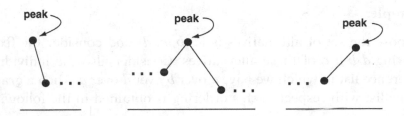

The following definition is the central concept in Black's theorem. It is what captures the notion of "coherence" that is exploited to yield the desired transitivity.

DEFINITION. A sequence of individual preference lists is said to satisfy *single peaked preferences* if there is a single ordering of the alternatives so that each of the individual preference lists has a graph—with respect to this ordering—that has only one peak.

Example

Consider the following three individual preference lists:

#1	#2	#3
a	c	b
b	a	a
c	b	c

We will show that this sequence of preference lists satisfies single peaked preferences. In order to do this, we must find a single ordering of the alternatives that will yield graphs of the individual preference lists that all have only one peak. Trial and error will work here, but for three people and three alternatives there is a method that will always produce such an ordering if one exists: choose an alternative that does not occur on the bottom of any of the individual preference lists, and put this alternative second in the ordering. In the present example, alternative *a* doesn't occur on the bottom of any of the lists, so we put it in the middle. Either b or c could go first; both will work. We'll choose the latter.

We claim that the ordering $c\ a\ b$ shows that the above sequence of individual preference lists satisfies single peaked preferences.

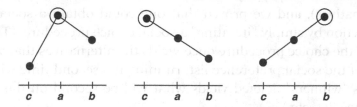

The single peak, in each case, has been circled. All three graphs could have been drawn above a single copy of the $c\ a\ b$ ordering.

At this point, we are ready to state and prove Black's theorem. Recall that the relation P is defined by xPy if more than half the people have x over y in their individual preference lists.

THEOREM. *(Black, 1958). Suppose that the number of individuals is odd and that we have a sequence of individual preference lists that satisfies single peaked preferences. Then the relation P is transitive.*

PROOF. Assume that we have a sequence of individual preference lists that satisfies single peaked preferences. We'll show that the sequence in Sen-coherent. Suppose that we have a triple of alternatives and assume (without loss of generality) that they occur in the order $x\,y\,z$ which demonstrates that single peaked preferences is satisfied. We must show that one of the three alternatives (x, y, z) fails to occur in one of the positions (above the other two, between the other two, or below the other two).

We claim that alternative y does not occur below the other two in any of the individual preference lists. If it did, there would be at least two peaks in the graph corresponding to that individual: one at the highest dot to the left of y, and one at the highest dot to the right of y.

The desired result now follows from Sen's theorem.

....

■ 7.7 CONCLUSIONS

We began this chapter by introducing the idea of a social welfare function (where the output is a social preference list instead of just a set of alternatives), and we proved that one could obtain a social welfare function by simply "iterating" a social choice procedure. (That is, running the choice procedure once yields the alternatives that will be on top of the social preference list; running it a second time with the previous "winners" deleted yields what will be second on the social preference list; etc.)

Taking the case of two alternatives as a starting point, we proved a generalized version of May's theorem for two alternatives that involved the idea of a quota of votes needed for passage. May's theorem itself—asserting that, for two alternatives and an odd number of people, majority vote is the only procedure that is anonymous, neutral, monotone, and does not produce ties—followed easily.

In **Section 7.4** we stated and proved Arrow's impossibility theorem, followed in **Section 7.5** with a proof of the Gibbard-Satterthwaite theorem. Finally, in **Section 7.6**, we considered the most natural relation (denoted here by P) that suggests itself for building a social preference list. (That is, say that aPb holds if more than half the people prefer a to b.) Coherency conditions on the sequence of individual preference lists yield situations where the relation P does, in fact, work. Two of the most well-known such results—those of Sen and Black—were presented in **Section 7.6**.

Exercises

1. Consider the following sequence of individual preference lists:

a	a	a	c	c	b	e
b	d	d	b	d	c	c
c	b	b	d	b	d	d
d	e	e	e	a	a	b
e	c	c	a	e	e	a

    (a) Consider the social welfare function arrived at by iterating the plurality procedure. Write down the social preference list that results from applying this function to the above sequence of individual preference lists.

(b) Do the same for the social welfare function arrived at by iterating the Borda count.

(c) Do the same for the social welfare function arrived at by iterating the Hare procedure.

(d) Do the same for the social welfare function arrived at by iterating sequential pairwise voting with the fixed agenda abcde.

(e) Do the same for the social welfare function arrived at by iterating the procedure where the last person on the right is a dictator.

2. Explain why:

(a) Procedure 1 in **Section 1.2** does not satisfy anonymity.

(b) Procedure 2 in **Section 1.2** does not satisfy neutrality.

(c) Procedure 3 in **Section 1.2** does not satisfy monotonicity.

3. Suppose $A = \{a, b, c\}$ and a given social welfare function produces:

$$
\begin{array}{ccc}
a & & c \quad a \quad b \\
\text{output} \quad b \quad \text{when confronted with input} & b \quad c \quad c \\
c & & a \quad b \quad a
\end{array}
$$

(a) If neutrality is satisfied, what is the output when confronted with input:

$$
\begin{array}{ccc}
a & c & b \\
b & a & a \\
c & b & c
\end{array}
$$

(b) What input would definitely yield c over a over b as output?

4. Prove that all quota systems satisfy anonymity, neutrality, and monotonicity.

5. In one or two sentences, explain why May's theorem follows from our result on quota systems.

6. Consider the social choice procedure that operates as follows. There are two fixed individuals (call them Person #1 and Person #2) and a fixed alternative (call it c). If both Person #1 and #2 are among the set of individuals, then Person #1 is the dictator if c is one of the alternatives, and Person #2 is the dictator if c is not one

of the alternatives. If either Person #1 or #2 is not among the set of individuals, then the procedure is a dictatorship with the person on the far left being the dictator.

(a) Show that this social choice procedure satisfies independence of irrelevant alternatives as defined in Chapter 1.

(b) Show that the iterated version of this procedure does not satisfy independence of irrelevant alternatives as defined in the present chapter by considering the following two sequences of individual preference lists:

Person #1	Person #2	Person #1	Person #2
c	b	a	b
a	a	b	a
b	c	c	c

7. Without using Arrow's theorem, show that if a social welfare function satisfies both Pareto and independence of irrelevant alternatives, then

(a) the set P of all individuals is a dictating set, and

(b) a set $\{p\}$ consisting of only one person is a dictating set if and only if p is a dictator.

8. Give two proofs showing that Lemma 5 implies Arrow's theorem— one based on splitting sets into two pieces of roughly equal size (intuitively: halving the set), and the other based on splitting a single element off the set.

9. Provide the argument that handles case 2 in Lemma 1 of Arrow's theorem.

10. Show that the assumption of monotonicity is not needed in Arrow's theorem by using independence of irrelevant alternatives and the Pareto condition to prove the following lemma:

LEMMA. Suppose that X has the property that for any two alternatives x and y, if everyone in X places x over y and everyone not in X places y over x then we get x over y in the social preference list. Suppose *IIA* and Pareto are satisfied. Then X is a dictating set.

11. If we had chosen to work in a context where ties are allowed in the social preference list, then Arrow's theorem no longer holds with the

conclusion that there is a person whose list is always identical to the social preference list. What can be concluded is the existence of a person that is a weak dictator in the sense that if this person places one alternative strictly above another, then the same is true in the social preference list. However, if this weak dictator places two alternatives tied, there is no guarantee they will be tied in the social preference list.

(a) Show that the proof given in this chapter is adequate to yield this version of Arrow's theorem with ties. (The conclusion will still be that there is a person p such that $\{p\}$ is a dictating set.)

(b) Consider the social welfare function that proceeds as follows: if Person #1 places x over y, then we get x over y in the social preference list: if Person #1 has x and y tied, then x and y appear in the social preference list as they do in the list of Person #2. (I.e., Person #2 gets to be the tie-breaker, if he chooses.) Show that this procedure satisfies all the conditions in the hypothesis of Arrow's theorem, but there is no (strong) dictator.

12. Prove that any social choice procedure satisfying down-monotonicity also satisfies monotonicity.

13. Give an example to show that each of the following social choice procedures are manipulable.

(a) Borda count

(b) Hare system

14. Prove that to show that a set of voters X can use alternative x to block alternative y, if suffices to produce a single sequence of individual preference lists in which the following all hold:

1. Everyone in X has x over y in his or her preference list.

2. Everyone not in X has y over x in his or her preference list.

3. The resulting social choice is x.

15. Prove Lemma 5 used in the proof of the Gibbard-Satterthwaite theorem.

16. Prove that Condorcet's method is non-manipulable if we allow the possibility that a social choice procedure produces no social choice.

17. Prove that if the social choice set always consists of exactly one alternative, then plurality is non-manipulable.

18. Prove that if we consider social choice procedures where ties in the outcome are allowed, then plurality is manipulable. Assume that if

a voter prefers candidate A to candidate B, then the outcome in which A alone is the social choice is preferable to the outcome in which A and B are tied for social choice.

19. We say that a social choice procedure is *group manipulable* if there is a collection of voters X and profiles B (which we think of as representing the sincere preference lists of all voters, including everyone in X) and B' (which we think of as representing the insincere preference lists of at least some of those voters in X) such that everyone not in X has the same preference list in both B and B', and all voters in X prefer the outcome with profile B' to the outcome with profile B.

    (a) Given an example to show that plurality is group-manipulable.

    (b) Prove that Condorcet's method is not group-manipulable if we allow the possibility that a social choice procedure produces no social choice.

20. Prove that the Gibbard-Satterthwaite theorem still holds if we allow ties in the individual preference lists.

21. Prove that the Gibbard-Satterthwaite theorem still holds if we replace the Pareto assumption with the strictly weaker condition of non-imposition (See Chapter 1, Exercise 31).

22. An *anti-dictatorship* is a social choice procedure in which the social choice is the unique alternative at the bottom of a fixed individual's preference list.

    (a) Prove that an anti-dictatorship satisfies the condition of non-imposition, but not Pareto.

    (b) Show that Arrow's theorem fails if Pareto is replaced by non-imposition.

23. Prove that any social choice procedure that is non-manipulable must also satisfy the monotonicity criterion.

24. In two or three paragraphs, discuss the differences between Arrow's theorem and the Gibbard-Satterthwaite theorem.

25. Write down an individual preference list with more than one peak.

26. Determine if the following sequence of five preference lists is Sen coherent.

a	a	c	d	b
c	b	a	b	d
b	d	b	a	a
d	c	d	c	c

27. Complete the following proof of the proposition preceding Sen's theorem. Assume that we have an odd number of people and the relation P is transitive. We want to show that we can arrange the alternatives in a list so that x is placed higher than y in our list precisely when xPy holds. The naïve way to proceed is to construct the list by simply inserting the alternatives one by one in such a way that if, for example, we are inserting y, then y is placed
    (a) below every alternative x that has already been placed and that satisfies xPy, and
    (b) above every alternative z that has already been placed and that satifies yPz.

    For this to work, it is necessary (and sufficient) to know that all the alternatives described in (a) have been placed higher in our list than all the alternatives described in (b). To see that this is true, take such an x and such a z ... (Reader: Take it from here ...)

28. A more concise proof of Sen's theorem can be based on the following lemma:

    LEMMA. *If $xPyPz$ holds, then someone ranks x above the other two, someone ranks y between the other two, and someone ranks z below the other two.*

    (a) Prove this lemma.
    (b) Use the lemma to prove Sen's theorem. (Hint: Assume that P is not transitive. Then there are alternatives a, b, and c so that $aPbPcPa$ holds. Notice that any of the alternatives a, b, or c can play the role of any one of the alternatives x, y, z in the lemma.

29. Provide the details in the argument for (a) case 4, (b) case 5, (c) case 6, (d) case 7, (e) case 8, (f) case 9 in the proof of Sen's theorem.

30. By considering the six distinct orderings of the three alternatives a, b, and c, show that the sequence of three preference lists used in the voting paradox does not satisfy single peaked preferences.

31. Show that the sequence of preference lists in Exercise 26 satisfies single peaked preferences.

32. The following exercise was suggested by William Zwicker, and and builds on Exercise 43 in Chapter 1. Consider the social welfare function arising from the Hare system in which the social preference list

ranks the alternatives in reverse order of elimination. This produces a social preference list that often has ties. The following modification in handling ties was proposed independently by V. Merlin and by V. Conitzer and M. Rognlie. They suggest adding a "fork" in the procedure when a tie occurs, making a separate copy of the profile for each of the alternatives involved in the tie, and for each copy of the profile, to continue with just that one alternative deleted at that stage. Of course, more forking might occur at later stages. The "winner" is a tie among the lists along each path in the tree created. Thus, this is not a social welfare function.

(a) Use the above procedure on the profiles P_1, P_2, and $P_1 + P_2$ from Exercise 43 in Chapter 1.

(b) Define what weak consistency in this context would mean and prove that this procedure is, in fact, weakly consistent.

CHAPTER

8

More Yes–No Voting

■ 8.1 INTRODUCTION

In this chapter, as in Chapter 2, our primary interest is in yes–no voting systems that are not weighted. We begin by returning to the theorem in Chapter 2 that characterized the weighted voting systems as precisely those that are trade robust (meaning that an arbitrary trade among several winning coalitions can never simultaneously render all of them losing). A natural question suggested by this result is whether trade robustness really needs to be stated in terms of "several winning coalitions." That is, perhaps a yes–no voting system is weighted if and only if a (not necessarily one-for-one) trade between *two* winning coalitions can never simultaneously render both losing. Recall that in showing that the procedure to amend the Canadian constitution is not trade robust we needed only two winning coalitions.

In **Section 8.2**, we show that life, or at least mathematics, is not quite that simple. There, we present a yes–no voting system that is not trade robust, but which has the property that an arbitrary trade between two winning coalitions always leaves at least one of them winning. Although not without its charms (we use a so-called magic square to construct the system), this is not a real-world example and, indeed, we know of no such real-world example.

In **Section 8.3**, we introduce the notion of the dimension of a yes–no voting system as the minimum number of weighted voting systems needed to realize the given system as their intersection. Thus, a yes–no voting system is of dimension 1 if and only it is weighted. We show that every yes–no voting system has a dimension, and the dimension of the United States federal system is (somewhat surprisingly) only 2. In **Section 8.4**, we generalize the notion of a weighted voting system by allowing the weights and quota to be vectors. We conclude by showing that every yes–no voting system is a "vector-weighted" system.

■ 8.2 A MAGIC SQUARE VOTING SYSTEM

Throughout the text, we have tried to use real-world examples whenever possible. Sometimes, however, one simply has to roll up one's sleeves and (mathematically) construct a voting system (or whatever) with the desired properties. This is precisely the situation we now face. That is, suppose we agree to call a yes–no voting system *two-trade robust* if an arbitrary exchange of players between two winning coalitions can never simultaneously render both losing. Thus, for example, the procedure to amend the Canadian Constitution is not two-trade robust. An affirmative answer to the following question would be nice in that it would yield a neater version of the characterization theorem in Chapter 2: Is every two-trade robust system automatically trade robust, and thus weighted? Unfortunately, the answer turns out to be "no," as shown by the following theorem from Taylor-Zwicker (1995a and 1999.):

> **THEOREM.** *There exists a yes–no voting system with nine voters that is two-trade robust but not trade robust, and thus not weighted.*

PROOF. Our starting point is the following array of nine numbers. It is called a "magic square" since the sum of every row and every column (as well as the diagonals) is the same (15).

$$4 \quad 3 \quad 8$$
$$9 \quad 5 \quad 1$$
$$2 \quad 7 \quad 6$$

We shall construct the desired yes–no voting system as follows: The voters will be the nine numbers $1, 2, \ldots, 9$. Every coalition with four or more voters will be winning and every coalition with two or fewer voters will be losing. For coalitions with exactly three voters, the ones with sum greater than 15 will be winning and the ones with sum less than 15 will be losing. Now, the only coalitions with exactly three voters and sum exactly equal to 15 are the rows, the columns, and the two diagonals. We shall declare the rows to be winning and the columns and diagonals to be losing.

We claim that this yes–no voting system is two-trade robust but not trade robust, and thus not weighted.

To see that the system is two-trade robust, suppose that we have two winning coalitions X and Y, and a trade between them yielding X' and Y'. We must show that either X' or Y' is still a winning coalition. We consider three cases:

Case 1: X or Y has Four or More Voters

Without loss of generality, assume it is X. Since Y is winning, Y has at least three voters. After the trade, the total number of voters is unchanged, and so at least one of X' and Y' has four or more voters and is thus winning, as desired.

Case 2: X or Y has Sum Strictly Greater than 15 and Each Contains Exactly Three Voters

Without loss of generality, assume X has sum greater than 15. Since Y is winning, Y has sum at least 15. Moreover, it is easy to see that the sum of X and Y is the same as the sum of X' and Y'. Thus, in this case, either X' has sum strictly greater than 15 (and is thus winning) or Y' has sum strictly greater than 15 (and is thus winning).

Case 3: X and Y have Sum Exactly Equal to 15 and Each Contains Exactly Three Voters

In this case X and Y must both be rows. If either X' or Y' has fewer than three voters, then the other will have more than three and so it will be winning, as desired. If either X' or Y' has sum less than 15, then the other will have sum greater than 15 and so it will be a winning coalition as desired. On the other hand, if both X' and Y' have sum exactly equal to 15 and exactly three voters, then the only way they could both be losing coalitions is if both were columns and/or diagonals. But one cannot convert two rows into two columns and/or diagonals by a trade (see Exercises 3 and 4). This completes the proof that the system is two-trade robust.

To see that the system is not weighted, it suffices (by the second theorem in **Section 2.5**) to show that it is not trade robust. For this, we begin with the the three rows $R1$, $R2$, and $R3$, which are winning coalitions. Thus,

$$R1 = \{4, 3, 8\};$$
$$R2 = \{9, 5, 1\};$$
$$R3 = \{2, 7, 6\}.$$

Now consider the trades that send

3 from $R1$ to $R2$,

8 from $R1$ to $R3$,

1 from $R2$ to $R3$,

9 from $R2$ to $R1$,

2 from $R3$ to $R1$, and

7 from $R3$ to $R2$.

These trades transform

$R1$ into $\{4, 9, 2\}$, which is the first column;

$R2$ into $\{3, 5, 7\}$, which is the second column; and

$R3$ into $\{8, 1, 6\}$, which is the third column.

Since the three columns are losing, this shows that the system is not trade robust. This completes the proof.

■ 8.3 DIMENSION THEORY AND THE U.S. FEDERAL SYSTEM

Our starting point in this section is the observation that although the procedure to amend the Canadian Constitution is not itself a weighted voting system, it is, in fact, constructed by "putting together" two weighted systems in a very natural way. That is, let's fix the set of voters to be the ten Canadian provinces and consider the following two yes–no voting systems:

System I will have as its winning coalitions precisely those consisting of seven or more voters (provinces). Let W_1 be this collection of coalitions.

System II will have as its winning coalitions precisely those representing at least half of Canada's population. Let W_2 be this collection of coalitions.

Notice, for example, that the coalition made up of the seven least populated provinces is in W_1 but not in W_2, while the coalition made up of the two most heavily populated provinces is in W_2 but not in W_1.

System I is a weighted voting system since we can give each of the provinces weight 1 and set the quota at 7. Similarly, System II is a weighted voting system since we can give each province weight equal to the percentage of Canada's population residing there and set the quota at 50.

Now, the procedure to amend the Canadian Constitution can be described by declaring a coalition to be winning if and only if it is winning in both System I and System II. That is, if we let W denote the collection of winning coalitions in the Canadian system, then a coalition X is in W if and only if it is in W_1 and W_2. Standard mathematical terminology would describe this by saying that the set W is the *intersection* of the sets W_1 and W_2. Notationally:

$$W = W_1 \cap W_2.$$

Thus, we have shown that the Canadian system is not a weighted voting system, but it can be described as the intersection of two weighted voting systems.

This discussion of the Canadian system suggests that one way to construct yes–no voting systems is to start with several different weighted voting systems all of which have the same set V of voters. One can then declare a coalition to be winning if and only if it is winning in all of these weighted systems. As we have seen, this allows one to construct non-weighted voting systems (like the Canadian system) from weighted ones. Remarkably, every yes–no voting system can be described in this way (without mention of vetoes or anything else). That is the content of the following proposition:

PROPOSITION. *Suppose S is a yes–no voting system for the set V of voters, and let m be the number of losing coalitions in S. Then it is possible to find m weighted voting systems with the same set V of voters such that a coalition is winning in S if and only if it is winning in every one of these m weighted systems. (Thus, the set of winning coalitions in S is the intersection of the sets of winning coalitions from these m weighted voting systems.)*

PROOF. For each losing coalition L in S, we construct a weighted voting system with set V of voters as follows: Every voter in L is given weight -1. Every voter not in L is given weight $+1$. The quota is set at $-|L|+1$, where $|L|$ denotes the number of voters in L. (For example, if L has 7 voters, then the quota is set at $-7+1=-6$.)

Notice that, in this weighted system, L is a losing coalition since it has weight $-|L|$ and $-|L| < -|L|+1$. However, every other coalition is a winning coalition in this weighted voting system, since they all have weight strictly larger than L does (that is, every other coalition has weight at least $-|L|+1$).

It now follows that if a coalition is winning in S, then it is winning in each of these weighted systems. Conversely, if a coalition is winning in each of these weighted systems, then it is a winning coalition in S. (That is, if it were a losing coalition in S, then it would be losing in the particular weighted system we built from that losing coalition.) Thus, a coalition is winning in S if and only if it is winning in each of these weighted systems. This completes the proof.

It is worth noting that if we start with a monotone yes–no voting system, then the above proof can be modified so that all the weights used are nonnegative (see Exercise 5).

The proof of the previous proposition provides a constructive procedure by which one can represent an arbitrary yes–no voting system as the intersection of weighted voting systems (in the sense that the set of winning coalitions in the original system is the set-theoretic intersection of the sets of winning coalitions from the weighted systems). This procedure, however, tends to be an enormously inefficient way to represent a given yes–no voting system as the intersection of weighted systems. For example, there are dozens of losing coalitions in the Canadian system, and so the proof of the above proposition would provide a representation of the Canadian system that is far less desirable than the representation we found of the Canadian system as the intersection of only two weighted systems. (On the other hand, Exercise 10 provides an example of a yes–no voting system wherein the representation provided by the proof of the above proposition is the best that can be achieved.)

Thus, although we have a proposition that guarantees every yes–no voting system can be represented as the intersection of weighted systems, it nevertheless becomes of interest to ask how efficiently this can be done for a given system. The question of efficiency leads to the following definition:

DEFINITION. A yes–no voting system is said to be of *dimension k* if and only if it can be represented as the intersection of exactly k weighted voting systems, but not as the intersection of $k - 1$ or fewer weighted voting systems.

Notice, for example, that a yes–no voting system is of dimension 1 if and only if it is weighted. We have already proved that the procedure to amend the Canadian constitution is of dimension 2. (It turns out that for each positive integer k, there is a voting system of dimension k—see Exercises 7, 8, and 9.)

As another illustration of how inefficient the representation given in the proof of the previous proposition can be, notice that the United States federal system has literally millions of losing coalitions. Moreover, although the House and the Senate function as weighted voting systems in their own right, the U.S. federal system is further complicated by the tie-breaking role of the vice president in the Senate, the veto power of the president, and the possibility of a Congressional override. Somewhat surprisingly, however, these emendations do not drive up the dimension of the U.S. federal system, as we now show.

PROPOSITION. *The U.S. federal system has dimension 2.*

PROOF. We know from Chapter 3 that the U.S. federal system is not weighted. Thus, it suffices to produce two weighted systems, with the same set of voters as the U.S. federal system, whose intersection is the U.S. federal system. The weighted systems that will do the trick are the following.

System I will give:

Weight 0 to each member of the House;
Weight 1 to each member of the Senate;
Weight $\frac{1}{2}$ to the vice president;
Weight $16\frac{1}{2}$ to the president;

and we set the quota at 67.

System II will give:

Weight 1 to each member of the House;
Weight 0 to each member of the Senate;
Weight 0 to the vice president;
Weight 72 to the president;

and we set the quota at 290.

We now want to show that a coalition is winning in the U.S. federal system if and only if it is winning in both System I and in System II. Suppose then that X is a coalition that is winning in the U.S. federal system. Without loss of generality, we can assume that X is a minimal winning coalition (Exercise 11 asks why we lose no generality with this assumption). Thus, X is one of the following three kinds of coalition:

1. X consists of 218 House Members, 51 senators, and the president;

2. X consists of 218 House Members, 50 senators, the vice president, and the president;

3. X consists of 290 House Members and 67 senators.

We leave it to the reader to verify that all three kinds of coalition achieve quota in both System I and in System II (see Exercise 12).

For the converse, assume that X is a winning coalition in both System I and in System II. We consider two cases:

Case 1: X Contains the President

Since X is winning in System I, it must have System I weight at least 67. Since the System I weight of the president is $16\frac{1}{2}$, the other members of X must contribute at least weight $50\frac{1}{2}$ to the total System I weight of X. But House members have weight 0 in System I, so X must contain either 51 (or more) senators or at least 50 senators and the vice president. Now, looking at the System II weight of X, which is at least 290 including the 72 contributed by the president, we see that X must also contain at least $290 - 72 = 218$ members of the House. Thus, in case 1, we see that X is a winning coalition in the federal system, as desired.

Case 2: X does not Contain the President

This is left to the reader (see Exercise 13), and completes the proof.

We conclude this section with the observation that we know of no real-world voting system of dimension 3 or higher.

■ 8.4 VECTOR-WEIGHTED VOTING SYSTEMS

In our early discussions of yes–no voting systems in Chapter 2, we suggested that the observation that the U.N. Security Council is, in fact, a weighted voting system might naturally lead one to conjecture that every yes–no voting system is weighted. We now know that not to be the case, and much of what we have done in Chapter 2 and Chapter 8 has been aimed at exploring the extent to which such a system can fail to be weighted. In this section, however, we show that the intuition provided by the weightedness of the U.N. Security Council is far less naïve than it might now seem.

Generalization has always played an important role in mathematics. For example, our original number system consisted of what we now call positive integers. This system was generalized to include zero, the negative numbers, then fractions, irrationals, and imaginaries. Of course, generalization for its own sake can at least sometimes be pointless. But a natural generalization of an important concept can often shed considerable light. Our goal in this section is to provide such a generalization of the notion of a weighted voting system.

Our starting point will be the observation that one can replace the notion of a real number by one of its generalizations: an ordered pair (x, y) of real numbers. These ordered pairs can be "added" as follows:

$$(x_1, y_1) + (x_2, y_2) = (x_1 + x_2, y_1 + y_2).$$

Thus, for example, $(2, 4) + (\frac{1}{2}, -1) = (\frac{5}{2}, 3)$. Moreover, we can "compare the size" of ordered pairs as follows:

$$(x_1, y_1) \leq (x_2, y) \text{ if and only if } x_1 \leq x_2 \text{ and } y_1 \leq y_2.$$

Now, let's return to the Canadian system (which we know is not weighted) and show that it is a "generalized weighted system." That is, instead of assigning real numbers as weights, let's assign ordered pairs as weights in the following way:

weight of Prince Edward Island will be (1, 0)
weight of Newfoundland will be (1, 2)
weight of New Brunswick will be (1, 2)
weight of Nova Scotia will be (1, 3)
weight of Manitoba will be (1, 4)
weight of Saskatchewan will be (1, 3)
weight of Alberta will be (1, 11)
weight of British Columbia will be (1, 13)
weight of Quebec will be (1, 23)
weight of Ontario will be (1, 39).

Notice that the first entry of each ordered pair is 1 and the second entry is the percentage of the Canadian population residing in that province. We shall let the ordered pair (7, 50) serve as the "quota."

Given a coalition, it now makes sense to define the weight of the coalition to be the ordered pair obtained by adding up all the ordered pair weights of the provinces in the coalition (just as we obtained the weight of a coalition in a weighted voting system by adding up the weights of all the voters in the coalition). This yields an ordered pair as "weight" for the coalition, which we can then compare (using $\leq$ as defined above) with the ordered pair that is the quota.

For example, if X is the coalition consisting of Manitoba, Saskatchewan, Alberta, British Columbia, and Ontario, then the "weight" of X is

$$(1, 4) + (1, 3) + (1, 11) + (1, 13) + (1, 39) = (5, 70).$$

If we compare (5, 70) with the quota (7, 50) we find that the weight of this coalition does not meet quota; that is, the statement

$$\text{"}(7, 50) \leq (5, 70)\text{"}$$

is not true since 7 is not less than or equal to 5.

Notice that with these definitions of "weight" and "quota," a coalition's weight meets quota if and only if it contains at least seven provinces (thus guaranteeing the first entry in its weight is at least as large as the first entry in the quota) and the combined population of the provinces in the coalition is at least half the Canadian population (thus guaranteeing that the second entry in its weight is at least as large as the second entry in the quota). Thus, a coalition meets quota if and only if it is a winning coalition in the Canadian system.

In the above discussion of the Canadian system, we used ordered pairs as the "weights" and "quota." As one might imagine, there are other examples where the weights and quota are ordered triples

$$(x, y, z)$$

that are "added" and "compared" in the obvious way. In general, if $x_1, x_2, \ldots, x_n$ are real number, then $(x_1, x_2, \ldots, x_n)$ is called an *ordered n-tuple*. Ordered n-tuples are added and compared as follows:

$$(x_1, x_2, \ldots, x_n) + (y_1, y_2, \ldots, y_n) = (x_1 + y_1, x_2 + y_2, \ldots, x_n + y_n),$$

and

$$(x_1, \ldots, x_n) \le (y_1, \ldots, y_n)$$

if and only if

$$x_1 \le y_1 \text{ and } \ldots \text{ and } x_n \le y_n.$$

All of this leads to the following definition:

DEFINITION. A yes–no voting system is said to be a *vector-weighted system* if, for some positive integer n, there exists an n-tuple "weight" for each voter and an n-tuple "quota" such that a coalition is winning precisely when the sum of the vector weights of the voters in the coalition meets or exceeds quota (in the sense of comparing two n-tuples described above).

Thus, for example, we have shown that the Canadian system is a vector-weighted system. Remarkably, the following turns out to be true.

THEOREM. *Every yes–no voting system is a vector weighted system. Moreover, if a system is of dimension n, then the weights and quota can be taken to be n-tuples but not $(n - 1)$-tuples.*

PROOF. Suppose $\mathcal{S}$ is an arbitrary yes–no voting system for the set V of voters. By the proposition in the last section, we know that $\mathcal{S}$ has dimension n for some n. Thus, we can choose weighted yes–no voting systems $\mathcal{S}_1, \ldots, \mathcal{S}_n$ so that for every coalition X from V, we have that

$$X \text{ is winning in } \mathcal{S}$$

if and only if

$$X \text{ is winning in } \mathcal{S}_1 \text{ and } \ldots \text{ and } X \text{ is winning in } \mathcal{S}_n.$$

To keep the notation simple, let's assume that $n = 3$. [Exercise 15(b) asks the reader to redo the proof using n in place of 3.]

Let w_1 and q_1 be the weight function and quota associated with $\mathcal{S}_1$, and similarly let w_2 and q_2, and w_3 and q_3 be those for $\mathcal{S}_2$ and $\mathcal{S}_3$ respectively. Thus, if X is a coalition, then

X is winning in $\mathcal{S}_1$ and X is winning in $\mathcal{S}_2$ and X is winning in $\mathcal{S}_3$

if and only if

$$w_1(X) \geq q_1 \text{ and } w_2(X) \geq q_2 \text{ and } w_3(X) \geq q_3.$$

If v is an arbitrary voter, we can produce a 3-tuple as weight for v by using the three weights he or she is assigned in the three weighted systems $\mathcal{S}_1$, $\mathcal{S}_2$, and $\mathcal{S}_3$ as follows:

$$w(v) = (w_1(v), w_2(v), w_3(v)).$$

Moreover, we can combine the three quotas q_1, q_2, and q_3 into a 3-tuple quota in the obvious way:

$$q = (q_1, q_2, q_3).$$

We must still show that these 3-tuple weights and quota "work" in the sense that a coalition should be winning in $\mathcal{S}$ if and only if its 3-tuple weight meets or exceeds quota (in the sense of comparing 3-tuples). Again, to keep the notation simple, let's assume we have a two-voter coalition $X = \{a, b\}$. Then

$$w_1(X) = w_1(a) + w_1(b);$$
$$w_2(X) = w_2(a) + w_2(b);$$
$$w_3(X) = w_3(a) + w_3(b).$$

Now, putting this together with what we had above yields

X is winning in $\mathcal{S}$

if and only if

X is winning in $\mathcal{S}_1$ and X is winning in $\mathcal{S}_2$ and X is winning in $\mathcal{S}_3$

if and only if

$$w_1(X) \geq q_1 \text{ and } w_2(x) \geq q_2 \text{ and } w_3(X) \geq q_3$$

if and only if

$$w_1(a) + w_1(b) \geq q_1 \text{ and } w_2(a) + w_2(b) \geq q_2 \text{ and } w_3(a) + w_3(b) \geq q_3$$

if and only if

$$(w_1(a) + w_1(b), w_2(a) + w_2(b), w_3(a) + w_3(b)) \geq (q_1, q_2, q_3)$$

if and only if

$$(w_1(a), w_2(a), w_3(a)) + (w_1(b), w_2(b), w_3(b)) \geq (q_1, q_2, q_3)$$

if and only if

$$w(a) + w(b) \geq q$$

if and only if

$$w(X) \geq q,$$

as desired.

Thus, the intuition that every yes–no voting system is a weighted system is justified—at least if one is willing to accept the naturality of *vector-valued* weights and quota.

....

■ 8.5 CONCLUSIONS

All the results in this chapter, as well as those in Chapter 3, were inspired by the following question:

Is every yes–no voting system weighted, and—if not—can we find a nice characterization of those that are?

It turned out that not every yes–no voting system is weighted, but that the weighted ones can be characterized as precisely those that are trade robust, meaning that an arbitrary sequence of trades among several winning coalitions can never render all of them losing.

Our starting point in this chapter was the question of whether we really needed to state trade-robustness in terms of several winning coalitions instead of just two winning coalitions. That is, perhaps a yes–no voting system is weighted if and only if one can never convert two winning coalitions into two losing coalitions by a trade. In **Section 8.2**, we showed that, alas, this is not the case. We constructed there a nine-person yes–no voting system (using a 3 by 3 magic square) that is not weighted but fails to be trade robust only when three or more coalitions are involved.

Section 8.3 returned to the attractive (but false) conjecture that every yes–no voting system is weighted. We showed there that, although the conjecture is not true, every yes–no voting system is the intersection of weighted systems. We then defined the dimension of a yes–no voting system to be the smallest number of weighted systems whose intersection is the given system. We also showed that the U.S. federal system—even with the tie-breaking vote of the vice president, the presidential veto, and the possibility of an override of such a veto by Congress—has dimension 2.

Section 8.4 revisited the material in Section 8.3 from a different point of view. We introduced the idea of a "vector-weighted" system as one in which the weights and quota are n-tuples of real numbers. The weight of a coalition could then be calculated by adding the n-tuple weights of its members, and this n-tuple weight of the coalition could then be compared with the n-tuple quota to see if the coalition "meets quota." We concluded our discussion of yes–no voting systems by proving that every yes–no voting system is a vector-weighted system.

EXERCISES

1. For the magic square yes–no voting system in **Section 8.2**, list out the minimal winning coalitions and the maximal losing coalitions.
2. Prove that the magic square yes–no voting system is monotone.
3. In the magic square yes–no voting system, prove that it is impossible to convert the first two rows into the first two columns by a trade. Prove that it is impossible to convert the first two rows into the first column and the upper right to lower left diagonal by a trade.
4. Generalize the result in Exercise 3 to explain why one can never convert two rows into two columns and/or diagonals by a trade.
5. Suppose S is a *monotone* yes–no voting system with set V of voters. Let $X_1, \ldots, X_n$ denote the *maximal* losing coalitions. For each $i = 1, \ldots, n$ let S_i be the weighted voting system that assigns weight 1 to each voter in X_i and weight $|X_i| + 1$ to each voter not in X_i (where $|X_i|$ denotes the number of voters in X_i). Find q so that X_i and subsets of X_i are the only losing coalitions in S_i and show that S is the intersection of these n *positively* weighted voting systems.

6. Prove or disprove: The dimension of a monotone yes–no voting system is always the same as the number of maximal losing coalitions.

7. Suppose we have four voters: A_1, A_2, A_3, and A_4. Let $\mathcal{S}_2$ denote the yes–no voting system wherein a coalition X is winning if and only if

   (i)   at least one of A_1 and A_2 is in X, and
   (ii) at least one of A_3 and A_4 is in X.

   Prove that $\mathcal{S}_2$ has dimension 2. (Hint: In one weighting, give A_1 and A_2 weight 3 and give A_3 and A_4 weight 1. Set the quota at 3. Do a similar thing for the other weighting.)

8. Suppose we have six voters: A_1, A_2, A_3, A_4, A_5, and A_6. Let $\mathcal{S}_3$ denote the yes–no voting system wherein a coalition X is winning if and only if

   (i)   at least one of A_1 and A_2 is in X,
   (ii)   at least one of A_3 and A_4 is in X, and
   (iii) at least one of A_5 and A_6 is in X.

   (a) Prove that $\mathcal{S}_3$ has dimension at most 3.
   (b) Suppose, for contradiction, that $\mathcal{S}_3$ can be expressed as the intersection of two weighted systems. Since $\{A_1, A_2, A_3, A_4\}$, $\{A_1, A_2, A_5, A_6\}$, and $\{A_3, A_4, A_5, A_6\}$ are all losing, each one must be losing in at least one of the two weighted systems. Hence, two of these—say $\{A_1, A_2, A_3, A_4\}$ and $\{A_3, A_4, A_5, A_6\}$—are losing in the same weighted system. Show that the desired contradiction now can be achieved by an appropriate trade.

9. Generalize Exercise 8 to show that for every n there is a monotone yes–no voting system of exact dimension n.

10. Suppose V is a set of n voters and let $\mathcal{S}$ be the yes–no voting system wherein a coalition is winning if and only if it contains an odd number of voters. Suppose we have a single weighted voting system for V with weight function w and quota q which makes all the coalitions with an odd number of voters winning. Suppose X and Y both have an even number of voters, v is in X but not in Y, $w(X) < q$, and $w(Y) < q$. Derive from this a contradiction, and conclude that the dimension of $\mathcal{S}$ is at least 2^{n-1}. (It turns out that there are

monotone yes–no voting systems with exponential dimension, but they are somewhat more elaborate than this one.)

11. Explain why, in the proof that the U.S. federal system has dimension 2, we lose no generality in working with *minimal* winning coalitions.

12. Verify that all three kinds of minimal winning coalitions in the U.S. federal system achieve quota in both System I and System II described in the proof that the U.S. federal system has dimension 2.

13. Give the argument omitted in the proof that the U.S. federal system has dimension 2.

14. In the system to amend the Canadian constitution, give three examples of minimal winning coalitions and three examples of maximal losing coalitions. For each of your six coalitions, calculate its vector-valued weight (which will be an ordered pair). In each case, compare the weight of the coalition with the quota (which is also an ordered pair) to verify the winningness or losingness of the coalition.

15. Prove that U.S. federal system is a vector weighted voting system by explicitly producing the weights and quota and demonstrating that they "work."

16. Redo the proof of the last theorem in this chapter
    (a) without the simplifying assumption that $|X| = 2$, and
    (b) without the simplifying assumptions that the dimension of the system is 3.

17. Consider the "minority veto" system where there are eleven voters, three of whom represent a minority interest. Passage requires at least six of the eleven votes subject to a veto by two or more of the three minority representatives. Show that this system is of dimension 2.

CHAPTER
9

More Political Power

■ 9.1 INTRODUCTION

We continue our study of political power in this chapter, beginning in **Sections 9.2** and **9.3** with two more quantitative measures of power. Both of these power indices were introduced in the late 1970s, the first appearing in Johnston (1978) and the second in Deegan–Packel (1978). These indices are similar in some ways to the Shapley–Shubik and Banzhaf indices introduced in Chapter 3, but they also differ in some important respects from these earlier ones as well as from each other. Additionally, we build on work of Brams, Affuso, and Kilgour (1989) in applying these indices to measure the power of the president in the context of the United States federal system as we did for the Shapley–Shubik index and the Banzhaf index in Chapter 3. It turns out, for example, that according to the Deegan–Packel index, the president has less than 1 percent of the power. The Johnston index, however, suggests that the president has 77 percent of the power.

In **Sections 9.4** and **9.5**, we offer a precise mathematical definition of what it means to say that two voters have comparable or incomparable power. This definition provides us with what is called an

ordinal notion of power—"ordinal" referring to the fact that the ordering is not derived from the assignment of numbers, as it was with the
four "cardinal" power indices (Shapley-Shubik, Banzhaf, Johnston,
and Deegan-Packel). This ordinal notion of power is closely linked
to the idea of "swap-robustness" from Chapter 2; the cognoscenti will
recognize it as leading to the well-known "desirability relation on individuals." We conclude in **Section 9.6** with a theorem from Straffin
(1980) that allows one to calculate the Shapley-Shubik index of a
so-called "voting bloc" in a fairly trivial way.

■ 9.2 THE JOHNSTON INDEX OF POWER

The Banzhaf index of power from Chapter 3 was based on the idea
of a critical defection from a winning coalition. It does not, however,
take into consideration the total number of players whose defection
from a given coalition is critical. The point is, one might well argue
that if a player p is the only one whose defection from C is critical,
then this is a stronger indication of power than if, say, every player
in C has a critical defection. This is the idea underlying the Johnston
index of power as formalized in the following two definitions (which
mimic those of **Section 3.4**).

DEFINITION. Suppose that p is one of the players in a yes–no voting
system. Then the *total Johnston power* of p, denoted here by TJP(p), is
the number arrived at as follows:

Suppose $C_1, \ldots, C_j$ are the winning coalitions for which p's defection
is critical. Suppose n_1 is the number of players whose defection from
C_1, is critical, n_2 is the number whose defection from C_2 is critical and
so on up to n_j being the number of players whose defection from C_j is
critical. Then

$$TJP(P) = \frac{1}{n_1} + \frac{1}{n_2} + \cdots + \frac{1}{n_j}.$$

DEFINITION. Suppose that p_1 is a player in a yes–no voting system and that the other players are denoted by $p_2, p_3, \ldots, p_n$. Then the
Johnston index of p_l, denoted here by $JI(p_l)$, is the number given by

$$JI(p_1) = \frac{TJP(p_1)}{TJP(p_1) + \cdots + TJP(p_n)}.$$

Example:

Let's stick with the same three player example that we used to illustrate both the Shapley–Shubik index and the Banzhaf index in Chapter 3. Thus, p_l has fifty votes, p_2 has forty-nine, and p_3 has one. Passage requires fifty-one votes. As before, the winning coalitions are

$$C_1 = \{p_l, p_2, p_3\}$$
$$C_2 = \{p_1, p_2\}$$
$$C_3 = \{p_1, p_3\}.$$

Now, p_1, has the only critical defection from C_1, but for both C_2 and C_3 it shares this property with the other member of the coalition. Thus, in calculating the absolute Johnston voting power of p_1, we get a contribution of 1 from p_1's presence in C_1, but only a contribution of $\frac{1}{2}$ each from its presence in C_2 and C_3. Similar comments hold for p_2 and p_3. Thus we have:

$$\text{TJP}(p_1) = 1 + \frac{1}{2} + \frac{1}{2} = 2$$

$$\text{TJP}(p_2) = 0 + \frac{1}{2} + 0 = \frac{1}{2}$$

$$\text{TJP}(p_3) = 0 + 0 + \frac{1}{2} = \frac{1}{2}$$

and

$$\text{JI}(p_1) = \frac{2}{2 + \frac{1}{2} + \frac{1}{2}} = \frac{2}{3}$$

$$\text{JI}(p_2) = \frac{\frac{1}{2}}{2 + \frac{1}{2} + \frac{1}{2}} = \frac{1}{6}$$

$$\text{JI}(p_3) = \frac{\frac{1}{2}}{2 + \frac{1}{2} + \frac{1}{2}} = \frac{1}{6}.$$

Notice that these turn out to agree with the Shapley–Shubik values in **Section 3.2**.

For yes–no voting systems where the number of winning coalitions is reasonably small, one can calculate total Johnston power by using a chart as we did for total Banzhaf power in **Section 3.6**. The difference is that, with Johnston power, one needs to identify which voters have critical defections from which winning coalitions. We illustrate this by calculating the total Johnston power for the European Economic Community as set up in 1958. Critical defections are italicized in the chart below, and the number of countries italicized in a winning coalition determines the fractions that occur to the right of that coalition. Thus, in the winning coalition listed below as *FIBN*L, defections by the four countries F, I, B, and N are critical. Hence, each of these four countries receives a contribution of $\frac{1}{4}$ from this winning coalition towards its total Johnston power.

	F	G	I	B	N	L
FGI	$\frac{1}{3}$	$\frac{1}{3}$	$\frac{1}{3}$			
FGBN	$\frac{1}{4}$	$\frac{1}{4}$		$\frac{1}{4}$	$\frac{1}{4}$	
FIBN	$\frac{1}{4}$		$\frac{1}{4}$	$\frac{1}{4}$	$\frac{1}{4}$	
GIBN		$\frac{1}{4}$	$\frac{1}{4}$	$\frac{1}{4}$	$\frac{1}{4}$	
FGIL	$\frac{1}{3}$	$\frac{1}{3}$	$\frac{1}{3}$			
FGBNL	$\frac{1}{4}$	$\frac{1}{4}$		$\frac{1}{4}$	$\frac{1}{4}$	
FIBNL	$\frac{1}{4}$		$\frac{1}{4}$	$\frac{1}{4}$	$\frac{1}{4}$	
GIBNL		$\frac{1}{4}$	$\frac{1}{4}$	$\frac{1}{4}$	$\frac{1}{4}$	
FGIB	$\frac{1}{3}$	$\frac{1}{3}$	$\frac{1}{3}$			
FGIN	$\frac{1}{3}$	$\frac{1}{3}$	$\frac{1}{3}$			
FGIBL	$\frac{1}{3}$	$\frac{1}{3}$	$\frac{1}{3}$			
FGINL	$\frac{1}{3}$	$\frac{1}{3}$	$\frac{1}{3}$			
FGIBN						
FGIBNL						
TJP	3	3	3	$\frac{3}{2}$	$\frac{3}{2}$	0
JI	$\frac{1}{4}$	$\frac{1}{4}$	$\frac{1}{4}$	$\frac{1}{8}$	$\frac{1}{8}$	0

Summarizing these results as we did for the previous power indices in Chapter 3 yields the following:

Country	Votes	Percentage of votes	JI	Percentage of power
France	4	23.5	$\frac{1}{4}$	25.0
Germany	4	23.5	$\frac{1}{4}$	25.0
Italy	4	23.5	$\frac{1}{4}$	25.0
Belgium	2	11.8	$\frac{1}{8}$	12.5
Netherlands	2	11.8	$\frac{1}{8}$	12.5
Luxembourg	1	5.9	0	0

The Johnston Index of the President

To calculate the Johnston index of the president, we need a break-down of the types of winning coalitions possible since we also must worry about how many players in each coalition have defections that are critical. In the chart that follows, we will give a name (like "T_{11}") for each type of winning coalition, and we will describe them by an expression like "67 S and 290 H" to indicate that this type of coalition is made up of 67 senators and 290 members of the House. "P" stands for "president."

Description of the winning coalitions	Number of critical defections	Whose defection is critical		
T_{11}: 67 S and 290 H	357		S	H
T_{12}: 67 S and 291–435 H	67		S	
T_{13}: 68–100 S and 290 H	290			H
T_{21}: P and 51 S and 218 H	270	P	S	H
T_{22}: P and 51 S and 219–289 H	52	P	S	
T_{23}: P and 52–66 S and 218 H	219	P		H
T_{24}: P and 52–66 S and 219–289 H	1	P		
T_{31}: P and 67–100 S and 218 H	219	P		H
T_{32}: P and 67–100 S and 219–289 H	1	P		
T_{41}: P and 51 S and 290–435 H	52	P	S	
T_{42}: P and 52–66 S and 290–435 H	1	P		

Now we can calculate the total Johnston power of the president, a member of the Senate, and a member of the House.

The winning coalitions involved in calculating the total Johnston power of the president can be obtained from column three above. They are

$$T_{21}, T_{22}, T_{23}, T_{24}, T_{31}, T_{32}, T_{41}, T_{42}.$$

Let $|T_{21}|$ denote the number of coalitions of type T_{21}. If we were calculating the total Banzhaf power of the president, we would just add the numbers: $|T_{21}|, |T_{22}|$, etc. However, since we are calculating the Johnston power, we must "adjust" each factor by dividing it by the number of people whose defection from such a coalition is critical. These can be obtained from column two above. Thus,

$$\text{TJP (The President)} = \frac{1}{270}|T_{21}| + \frac{1}{52}|T_{22}| + \frac{1}{219}|T_{23}|$$
$$+ |T_{24}| + \frac{1}{219}|T_{31}| + |T_{32}| + \frac{1}{52}|T_{41}| + |T_{42}|.$$

The calculation of, say, $|T_{24}|$, proceeds in a way similar to what we did in the calculations of the Banzhaf power. That is,

$$|T_{24}| = \left[\binom{100}{52} + \cdots + \binom{100}{66} \right] \times \left[\binom{435}{219} + \cdots + \binom{435}{289} \right].$$

Expressing $|T_{21}|, |T_{23}|$, etc. in "n choose k" notation is left to the reader. (See Exercises 5 and 6 at the end of the chapter.)

Now let's consider a fixed member of the Senate. Column three of the previous chart again shows which types of coalitions will have to be considered. They are

$$T_{11}, T_{12}, T_{21}, T_{22}, T_{41}.$$

Here, however, we have to be a little careful since we do not want to use, for example, $|T_{11}|$. The point is, lots of the coalitions of type T_{11}, do not even include as a member the particular senator we are considering. Thus, the contribution from type T_{11} coalitions will involve the number of ways of choosing 66 other senators from the pool of 99 remaining senators multiplied by the number of ways of choosing 290 members of the House from the available 435. In particular, the "n choose k" expression will involve

$$\binom{99}{66} \text{ and not } \binom{100}{67}$$

With this potential pitfall confronted, the calculations proceed in a way similar to what we've done before, yielding

$$\text{TJP(A Senator)} = \frac{1}{357}\left[\binom{99}{66} \times \binom{435}{290}\right]$$

$$+ \frac{1}{67}\binom{99}{66}\left[\binom{435}{291} + \cdots + \binom{435}{435}\right]$$

$$+ \frac{1}{270}\left[\binom{99}{50} \times \binom{435}{218}\right]$$

$$+ \frac{1}{52}\binom{99}{50}\left[\binom{435}{219} + \cdots + \binom{435}{289}\right]$$

$$+ \frac{1}{52}\binom{99}{50}\left[\binom{435}{290} + \cdots + \binom{435}{435}\right].$$

Expressing the total Johnston power of a member of the House of Representatives in a similar fashion is left to the reader. (See Exercise 8 at the end of the chapter.)

To obtain the desired Johnston indices, we sum the total Johnston power of the 536 players involved, and then divide by the total. The results turn out to be:

$$\begin{aligned}
\text{JI(The president)} & = .77 \\
\text{JI(A senator)} & = .0016 \\
\text{JI(A member of the House)} & = .00017.
\end{aligned}$$

Expressing these in terms of percentages of power instead of small decimals yields:

(Johnston) Power held by the president $= 77\%$

(Johnston) Power held by the Senate $= 16\%$

(Johnston) Power held by the House $= 7\%$

The striking thing to notice is the very different measure of power assigned the president by this index as opposed to those in Chapter 3. A little more on this will be said in the concluding section of this chapter, but there is no substitute for going directly to the literature.

··

■ 9.3 THE DEEGAN–PACKEL INDEX OF POWER

In 1978, Deegan and Packel introduced a power index based on three assumptions:

1. Only minimal winning coalitions should be considered when determining the relative power of voters.

2. All minimal winning coalitions form with equal probability.

3. The amount or power a player derives from belonging to some minimal winning coalition is the same as that derived by any other player belonging to that same minimal winning coalition.

These assumptions, in fact, uniquely determine a power index. Moreover, the calculation required involves a nice blend of what we did with the two procedures for Banzhaf power (**Section 3.5**) and the calculation of Johnston power (**Section 9.2**). We begin with two definitions.

DEFINITION. Suppose that p is one of the voters in a yes–voting system. Then the *total Deegan–Packel power* of p, denoted here by TDPP(p), is the number arrived at as follows:

Suppose $C_1, \ldots, C_j$ are the *minimal* winning coalitions to which p belongs. Suppose n_1 is the number of voters in C_1, n_2 is the number in C_2, and so on up to n_j being the number of voters in C_j. Then

$$\text{TDPP}(p) = \frac{1}{n_1} + \frac{1}{n_2} + \cdots + \frac{1}{n_j}.$$

DEFINITION. Suppose that p_1 is a voter in a yes–no voting system and that the other voters are denoted by $p_2, p_3, \ldots, p_n$. Then the *Deegan–Packel index* of p_1, denoted here by DPI(p_1), is the number given by

$$\text{DPI}(p_1) = \frac{\text{TDPP}(p_1)}{\text{TDPP}(p_1) + \cdots + \text{TDPP}(p_n)}.$$

Example:

Suppose again that p_1 has fifty votes, p_2 has forty-nine votes, and p_3 has one vote, with passage requiring fifty-one votes. The minimal winning coalitions (subscripted as before) are:

$$C_2 = \{p_1, p_2\}$$
$$C_3 = \{p_1, p_3\}.$$

In calculating total Deegan–Packel power, p_1 receives a contribution of $\frac{1}{2}$ from each of the two minimal winning coalitions, while p_2 and p_3 each receive such a contribution from only one of the two minimal winning coalitions. Thus:

$$\text{TDPP}(p_1) = \frac{1}{2} + \frac{1}{2} = 1$$
$$\text{TDPP}(p_2) = \frac{1}{2} + 0 = \frac{1}{2}$$
$$\text{TDPP}(p_3) = 0 + \frac{1}{2} = \frac{1}{2}$$

and

$$\text{DPI}(p_1) = \frac{1}{1 + \frac{1}{2} + \frac{1}{2}} = \frac{1}{2}$$
$$\text{DPI}(p_2) = \frac{\frac{1}{2}}{1 + \frac{1}{2} + \frac{1}{2}} = \frac{1}{4}$$
$$\text{DPI}(p_3) = \frac{\frac{1}{2}}{1 + \frac{1}{2} + \frac{1}{2}} = \frac{1}{4}.$$

For an additional illustration of how to calculate Deegan–Packel power, we will return to the European Economic Community as set up in 1958. Notice that the chart we use now only includes the *minimal winning coalitions* in the left hand column.

	F	G	I	B	N	L
FGI	$\frac{1}{3}$	$\frac{1}{3}$	$\frac{1}{3}$			
FGBN	$\frac{1}{4}$	$\frac{1}{4}$		$\frac{1}{4}$	$\frac{1}{4}$	
FIBN	$\frac{1}{4}$		$\frac{1}{4}$	$\frac{1}{4}$	$\frac{1}{4}$	
GIBN		$\frac{1}{4}$	$\frac{1}{4}$	$\frac{1}{4}$	$\frac{1}{4}$	
TDPP	$\frac{5}{6}$	$\frac{5}{6}$	$\frac{5}{6}$	$\frac{3}{4}$	$\frac{3}{4}$	0
DPI	$\frac{5}{24}$	$\frac{5}{24}$	$\frac{5}{24}$	$\frac{3}{16}$	$\frac{3}{16}$	0

Summarizing these results as we did for the previous power indices yields the following:

Country	Votes	Percentage of votes	DPI	Percentage of power
France	4	23.5	$\frac{5}{24}$	20.8
Germany	4	23.5	$\frac{5}{24}$	20.8
Italy	4	23.5	$\frac{5}{24}$	20.8
Belgium	2	11.8	$\frac{3}{16}$	18.8
Netherlands	2	11.8	$\frac{3}{16}$	18.8
Luxembourg	1	5.9	0	0

The Deegan–Packel Index of the President

Any minimal winning coalition in the federal system that contains the president can be constructed by first choosing 51 members of the Senate and then choosing 218 members of the House. Hence, the total number of such minimal winning coalitions is

$$A = \binom{100}{51}\binom{435}{218}.$$

Similarly the total number of minimal winning coalitions that do not contain the president is given by

$$B = \binom{100}{67}\binom{435}{290}.$$

Note that every minimal winning coalition of the first type contains 270 voters (and so will contribute $\frac{1}{270}$ to the total Deegan–Packel power of each of its members), and every minimal winning coalition of the second type contains 357 voters (and so will contribute $\frac{1}{357}$ to the total Deegan–Packel power of each of its members).

It follows from the above that the number of minimal winning coalitions is $A + B$ (and so we will be dividing by $A + B$ in passing from total Deegan–Packel power to the Deegan–Packel index of each player). First, however, we note that we immediately have the following:

$$\text{TDPP(president)} = \frac{A}{270}.$$

It also turns out (see Exercise 14) that

$$\text{TDPP(A senator)} = \frac{1}{357}\binom{99}{66}\binom{435}{290} + \frac{1}{270}\binom{99}{50}\binom{435}{218}$$

and

$$\text{TDPP(A House member)} = \frac{1}{357}\binom{100}{67}\binom{434}{289} + \frac{1}{270}\binom{100}{51}\binom{434}{217}.$$

Dividing each of these expressions by $A + B$ (and using *Mathematica* to do the calculations) yields:

$$
\begin{aligned}
&\text{DPI(the president)} &&= .0037 \\
&\text{DPI(a senator)} &&= .0019 \\
&\text{DPI(a member of the House)} &&= .0019
\end{aligned}
$$

Again expressing these in terms of percentage of power instead of small decimals, we have:

(Deegan–Packel) Power held by the president $= .4\%$
(Deegan–Packel) Power held by the Senate $= 18.9\%$
(Deegan–Packel) Power held by the House $= 80.7\%$

For more on the Deegan–Packel index and the U.S. federal system, see Packel (1981).

....

■ 9.4 ORDINAL POWER: INCOMPARABILITY

As we did in Chapter 3, we will assume throughout this section that "yes–no voting system" means "*monotone* yes–no voting system." Thus, winning coalitions remain winning if new voters join them.

Suppose we have a yes–no voting system (and, again, not necessarily a weighted one) and two voters whom we shall call x and y. Our starting point will be an attempt to formalize (that is, to give a rigorous mathematical definition for) the intuitive notion that underlies expressions such as the following:

"x and y have equal power"

"x and y have the same amount of influence"

"x and y are equally desirable in terms of the formation of a winning coalition"

The third phrase is most suggestive of where we are heading and, in fact, the thing we are leading up to is widely referred to as the "desirability relation on individuals" (although we could equally well call it the "power ordering on individuals" or the "influence ordering on individuals"). We shall begin with an attempt to formalize the notion of x and y having "equal influence" or being "equally desirable."

If we think of the desirability of x and of y to a coalition Z, then there are four types of coalitions to consider:

1. x and y both belong to Z.

2. x belongs to Z but y does not.

3. y belongs to Z but x does not.

4. Neither x nor y belongs to Z.

If x and y are equally desirable (to the voters in Z, who want the coalition Z to be a winning one), then for each of the four situations described above, we have a statement that should be true:

1. If Z is a winning coalition, then x's defection from Z should render it losing if and only if y's defection from Z renders it losing.

2. If x leaves Z and y joins Z, then Z should go neither from being winning to being losing nor from being losing to being winning.

3. If y leaves Z and x joins Z, then Z should go neither from being winning to being losing nor from being losing to being winning.

4. x's joining Z makes Z winning if and only if y's joining Z makes Z winning.

In fact, it turns out that condition 4 is strong enough to imply the other three (see Exercises 11 and 12). This leads to the following definition:

DEFINITION. Suppose x and y are two voters in a yes–no voting system. Then we shall say that x and y are *equally desirable* (or, the desirability of x and y is equal, or the same), denoted $x \approx y$, if and only if the following holds:

For every coalition Z containing neither x nor y,
the result of x joining Z is a winning coalition
if and only if
the result of y joining Z is a winning coalition.

For brevity, we shall sometimes just say: "x and y are equivalent" when $x \approx y$.

Example:

Consider again the weighted voting system with three players a, b, and c who have weights 1, 49, and 50, respectively, and with quota $q = 51$. Then the winning coalitions are $\{a,c\}$, $\{b,c\}$, and $\{a,b,c\}$. Notice that $a \approx b$: the only coalitions containing neither a nor b are the empty coalition (call it Z_1) and the coalition consisting of c alone (call it Z_2). The result of a joining Z_1 is the same as the result of b joining Z_1 (a losing coalition) and the result of a joining Z_2 is the same as the result of b joining Z_2 (a winning coalition). On the other hand, a and c are not equivalent, since neither belongs to $Z = \{b\}$, but a joining Z yields $\{a, b\}$ which is losing with 50 votes, while c joining Z yields $\{b, c\}$ which is winning with 51 votes.

This example shows that in a weighted voting system, two voters with very different weights can be equivalent and, thus (intuitively) have the same "power" or "influence."

The relation of "equal desirability" defined above will be further explored in **Section 9.5**. For now, however, we turn our attention to the question of when two voters not only fail to have equal influence, but when it makes sense to say that their influence is "incomparable." What should this mean? Mimicking what we did for the notion of equal desirability, let's say that x and y are incomparable if one coalition Z desires x more than y, and another coalition Z' desires y more than x. Formalizing this yields:

DEFINITION. For two voters x and y in a yes–no voting system, we say that the *desirability of x and y is incomparable*, denoted

$$x \mathbin{I} y$$

if and only if there are coalitions Z and Z', neither one of which contains x or y, such that the following hold:

1. the result of x joining Z is a winning coalition, but the result of y joining Z is a losing coalition, and

2. the result of y joining Z' is a winning coalition, but the result of x joining Z' is a losing coalition.

For brevity, we shall sometimes just say "x and y are incomparable" when $x \mathbin{I} y$.

Example:

In the U.S. federal system, let x be a member of the House and let y be a member of the Senate. Then $x \mathbin{I} y$ (see Exercise 14). On the other hand if x is the vice president and y is a member of the Senate, then x and y are not incomparable (see Exercise 15).

The following proposition characterizes exactly which yes–no voting systems will have incomparable voters. Recall from **Section 2.4** that a yes–no voting system is swap robust if a one-for-one exchange of players between two winning coalitions always leaves at least one of the two coalitions winning.

PROPOSITION. *For any yes–no voting system, the following are equivalent:*

    1.   There exist voters x and y whose desirability is incomparable.

    2.   The system fails to be swap robust.

PROOF. (1 implies 2): Assume that the desirability of x and y is incomparable, and let Z and Z' be coalitions such that:

    Z with x added is winning;
    Z with y added is losing;
    Z' with y added is winning; and
    Z' with x added is losing.

To see that the system is not swap robust, let X be the result of adding x to the coalition Z, and let Y be the result of adding y to the coalition Z'. Both X and Y are winning, but the one-for-one swap of x for y renders both coalitions losing.

(2 implies 1): Assume the system is not swap robust. Then we can choose winning coalitions X and Y with x in X but not in Y, and y in Y but not in X, such that both coalitions become losing if x is swapped for y. Let Z be the result of deleting x from the coalition X, and let Z' be the result of deleting y from the coalition Y. Then

    Z with x added is X, and this is winning;
    Z with y added is losing;
    Z' with y added is Y, and this is winning; and
    Z' with x added is losing.

This shows that the desirability of x and y is incomparable and completes the proof.

COROLLARY. *In a weighted voting system, there are never voters whose desirability is incomparable.*

PROOF. In **Section 2.4**, we showed that a weighted voting system is always swap robust.

The question of what one can say about voters x and y whose desirability is neither equal nor incomparable is taken up next, but, in the meantime, the reader can try Exercise 17.

....

■ 9.5 ORDINAL POWER: COMPARABILITY

The emphasis in **Section 9.4** was on formalizing the idea of what it means to say that two voters in a yes–no voting system have incomparable power. Here, however, we switch our emphasis to the question of how we can use ordinal notions to formalize the idea of two voters having comparable power.

The binary relation of "equal desirability" (**Section 9.4**) turns out to be what is called an *equivalence relation.* This means that the relation is reflexive, symmetric, and transitive. These notions are defined in the course of recording the following proposition:

PROPOSITION. *The relation of equal desirability is an equivalence relation on the set of voters in a yes–no voting system. That is, the following all hold:*

1. *The relation is reflexive: if $x = y$ (that is, if x and y are literally the same voter), then x and y are equally desirable.*

2. *The relation is symmetric: if x and y are equally desirable, then y and x are equally desirable.*

3. *The relation is transitive: if x and y are equally desirable and y and z are equally desirable, then x and z are equally desirable.*

REMARK. The reader should avoid letting our use of the phrase *equally desirable* lull him or her into thinking that the theorem is obvious. The only thing that is obvious is that if the theorem could not be rigorously established using the precise formal definition of *equal desirability* that we gave, then we would have been way out of line in choosing this phrase (loaded as it is with intuition) for the mathematical notion presented in the previous definition.

PROOF. We leave 1 and 2 to the reader (see Exercise 15). For 3, assume that x and y are equally desirable and that y and z are equally desirable. We want to show that x and z are equally desirable. Assume then that Z is an arbitrary coalition containing neither x nor z. We must show that the result of x joining Z is a winning coalition if and only if the result of z joining Z is a winning coalition. We consider two cases:

Case 1: y Does not Belong to Z

Since $x \approx y$ and neither x nor y belongs to Z, we know that the result of x joining Z is a winning coalition if and only if the result of y joining Z is a winning coalition. But now, since $y \approx z$ and neither y nor z belongs to Z, we know that the result of y joining Z is a winning coalition if and only if the result of z joining Z is a winning coalition. Thus, the result of x joining Z is a winning coalition if and only if the result of z joining Z is a winning coalition, as desired.

Case 2: y Belongs to Z

This case is quite a bit more difficult than the last one, and the reader should expect to spend several minutes checking to see that each line of the proof follows from previous lines.

We will make use of some set-theoretic notation in what follows. Suppose C is a coalition and v is a voter. Then

1. $C \cup \{v\}$ denotes the coalition resulting from v joining C. Typically, this is used when v does not already belong to C. If v does belong to C, then $C \cup \{v\}$ is the same as C.

2. $C - \{v\}$ denotes the coalition resulting from v leaving C. Typically, this is used when v belongs to C. If v does not belong to C, then $C - \{v\}$ is the same as C.

With this notation at hand, we can proceed with case 2.

Let A denote the coalition resulting from y leaving Z. Thus

$$A = Z - \{y\}$$

and so

$$Z = A \cup \{y\}.$$

Assume that $Z \cup \{x\}$ is a winning coalition. We want to show that $Z \cup \{z\}$ is also a winning coalition. Now,

$$Z \cup \{x\} = A \cup \{y\} \cup \{x\} = A \cup \{x\} \cup \{y\}.$$

Let $Z' = A \cup \{x\}$. Thus $Z' \cup \{y\}$ is a winning coalition. Since $y \approx z$ and neither y nor z belongs to Z', we know that $Z' \cup \{z\}$ is also a winning coalition. But $Z' \cup \{z\} = A \cup \{x\} \cup \{z\} = A \cup \{z\} \cup \{x\}$. Let $Z'' = A \cup \{z\}$. Thus $Z'' \cup \{x\}$ is a winning coalition. Since $x \approx y$ and neither x nor y belongs to Z'', we know that $Z'' \cup \{y\}$ is also a winning coalition. But $Z'' \cup \{y\} = A \cup \{z\} \cup \{y\} = A \cup \{y\} \cup \{z\} = Z \cup \{z\}$. Thus, $Z \cup \{z\}$ is a winning coalition as desired.

A completely analogous argument would show that if $Z \cup \{z\}$ is a winning coalition, then so is $Z \cup \{x\}$. This completes the proof.

For weighted voting systems, a naïve intuition would suggest that x and y are equally desirable precisely when they have the same weight. The problem with this intuition is that a given weighted voting system can be equipped with weights in many different ways. For example, consider the yes–no voting system corresponding to majority rule among three voters. This is a weighted voting system, as can be seen by assigning each of the voters weight 1 and setting the quota at 2. But the same yes–no voting system is realized by assigning the voters weights 1, 100, and 100, and setting the quota at 101. Notice that all three voters have the same weight in one of the weighted systems, but not in the other.

The above intuition, however, is not that far off. In fact, for a weighted voting system, we can characterize exactly when two voters are equally desirable as follows:

PROPOSITION. *For any two voters x and y in a weighted voting system, the following are equivalent:*

1. *x and y are equally desirable.*

2. *There exists an assignment of weights to the voters and a quota that realize the system and that give x and y the same weight.*

3. *There are two different ways to assign weights to the voters and two (perhaps equal) quotas such that both realize the system, but in one of the two weightings, x has more weight than y and, in the other weighting, y has more weight than x.*

PROOF. We first prove that 1 implies 2. Assume that x and y are equally desirable and choose any weighting and quota that realize the system. Let $w(x)$, $w(y)$ and q denote (respectively) the weight of x, the weight of y, and the quota. Expressions like "$w(Z)$" will represent the total weight of the coalition Z. We now construct a new weighting (where we will use nw for "new weight" in place of w for "weight") such that, with the same quota q, this new weighting also realizes the system and $nw(x) = nw(y)$.

The new weighting is obtained by keeping the weight of every voter except x and y the same, and setting both $nw(x)$ and $nw(y)$ equal to the average of $w(x)$ and $w(y)$.

To see that this new weighting still realizes the same system, assume that Z is a coalition. We must show that Z is winning in the new weighting if and only if Z is winning in the old weighting. We consider three cases:

Case 1: Neither x nor y Belongs to Z

In this case, $w(Z) = nw(Z)$ and so $nw(Z) \geq q$ if and only if Z is winning.

Case 2: Both x and y Belong to Z

We leave this for the reader.

Case 3: x Belongs to Z but y Does not Belong to Z

In this case, the new weight of Z is the average of the old weight of Z and the old weight of $Z - \{x\} \cup \{y\}$. That is:

$$nw(Z) = \frac{w(Z) + w(Z - \{x\} \cup \{y\})}{2}$$

Since x and y are equally desirable, either both Z and $Z - \{x\} \cup \{y\}$ are winning or both are losing. If both are winning, then

$$nw(Z) \geq \frac{q+q}{2} = q.$$

If both are losing, then

$$nw(Z) < \frac{q+q}{2} = q.$$

This completes the proof that 1 implies 2.

We now prove that 2 implies 3. Assume that we start with a weighting and quota wherein x and y have the same weight. Let HL denote the weight of the heaviest losing coalition, and let LW denote the weight of the lightest winning coalition. Thus,

$$HL < q \leq LW.$$

Let q' be the average of HL and q. Then q' still works as a quota and

$$HL < q' < LW.$$

Let ϵ be any positive number that is small enough so that

$$HL + \epsilon < q' < LW - \epsilon.$$

We leave it for the reader to check that the system is unchanged if we either increase the weight of x by ϵ or decrease the weight of x by ϵ. This shows that there are two weightings that realize the system, one of which makes x heavier than y and the other of which makes y heavier than x.

Finally, we prove that 3 implies 1. Assume that we have two weightings, w and w', and two quotas, q and q', such that

1. A coalition Z is winning if and only if $w(Z) \geq q$.

2. A coalition Z is winning if and only if $w'(Z) \geq q'$.

3. $w(x) > w(y)$.

4. $w'(y) > w'(x)$.

To show that x and y are equally desirable, we must start with an arbitrary coalition Z containing neither x nor y and show that $Z \cup \{x\}$ is winning if and only if $Z \cup \{y\}$ is winning. This argument is asked for in Exercise 20. Given this, the proof is complete.

Finally, what can we say about voters x and y whose desirability is neither equal nor incomparable? Looking back at the definition, we see that this happens only if (intuitively) some coalition desires one more than the other, but no coalition desires the other more than this one. Formally:

DEFINITION. For any two voters x and y in a yes–no voting system, we say that x *is more desirable than* y, denoted

$$x > y,$$

if and only if the following hold:

1. for every coalition Z containing neither x nor y, if $Z \cup \{y\}$ is winning then so is $Z \cup \{x\}$, and

2. there exists a coalition Z' containing neither x nor y such that $Z' \cup \{x\}$ is winning, but $Z' \cup \{y\}$ is losing.

We shall also write $x \geq y$ to mean that either $x > y$ or $x \approx y$. (This is analogous to what is done with numbers.) The relation $\geq$ is known in the literature as the *desirability relation on individuals*.

Example:

In the U.S. federal system if x is a senator and y is the vice president, then $x > y$ (see Exercise 21).

The binary relation $\geq$ is called a *preordering* because it is transitive and reflexive. A preordering is said to be *linear* if for every x and y one has either $x \geq y$ or $y \geq x$. Linear preorders are also called *weak orderings* in the literature. With this, we conclude the present discussion with one more definition and one more proposition.

DEFINITION. A yes–no voting system is said to be *linear* if there are no incomparable voters (equivalently, if the desirability relation on individuals is a linear preordering).

PROPOSITION. *A yes–no voting system is linear if and only if it is swap robust.*

COROLLARY. *Every weighted voting system is linear.*

For proofs of these, see Exercise 22.

Finally, for weighted voting systems, we have the following very nice characterization of the desirability relation on individuals.

PROPOSITION. *In a weighted voting system we have $x > y$ if and only if x has strictly more weight than y in every weighting that realizes the system.*

A proof of this (which is quite short, given what we did earlier in this section) is asked for in Exercise 23.

■ 9.6 A THEOREM ON VOTING BLOCS

This section considers a situation that reduces to a kind of weighted voting body that is sufficiently simple so that we can prove a general theorem, taken from Straffin (1980), that allows us to calculate the Shapley–Shubik indices of the players involved in an easy way. We begin with some notation and an example.

NOTATION. Suppose we have a weighted voting system with n players $p_1, \ldots, p_n$ with weights $w_1, \ldots, w_n$ (so, w_1 is the weight of player p_1, w_2 of p_2, etc.) Suppose that q is the quota. Then all of this is denoted by:

$$[q : w_1, w_2, \ldots, w_n].$$

For example, the European Economic Community is the system:

$$[12 : 4, 4, 4, 2, 2, 1]$$

With this notation at hand, we now turn to an extended example of how the Shapley–Shubik index of a voting bloc can be calculated.

Extended Example:

Consider the United States Senate as a yes–no voting system with 100 voters, each of whom has one vote and with fifty-one votes needed for passage. (We ignore the vice president.) Thus, the Shapley–Shubik index of any one senator is 1/100, since they must all be the same and sum to one. But now suppose the twelve senators from the six New England states decide to vote together as a so-called voting bloc. Intuitively, the power of this bloc would seem to be greater than the sum of the powers of the individual senators. Our goal in this example is to treat the bloc as a single player with twelve votes, and to calculate the Shapley–Shubik index of this bloc. The result obtained should quantify the above intuition.

Consider, then, the weighted voting body $[51 : 12, 1, 1, \ldots, 1]$ where there are eighty-eight ones (representing the senators from the forty-four non-New England states). The first thing to notice is that we really don't have to consider all possible ways of ordering the eighty–nine players. That is, this collection of 89! orderings is broken into 89 equal size "clumps" determined by the place occupied by the "12" in the string of eighty-eight ones. The eighty-nine "places" are pictured below.

$$\underline{\qquad} \; \overset{1}{\underline{\qquad}} \; \overset{1}{\cdots} \; \overset{1}{\underline{\qquad}} \; \overset{1}{\underline{\qquad}}$$
place 1 place 2   place 88   place 89
(88 ones all together)

The different orderings within any one clump are arrived at simply by permuting the "ones" involved. In particular, then, the twelve-vote player is pivotal for one ordering in the clump only if it is pivotal for every ordering in the clump. Thus, to calculate the Shapley–Shubik index of the voting bloc, we must simply determine how many of the eighty-nine distinct orderings (one from each clump) have the "12" in a pivotal position.

In order for the "12" to be in the pivotal position, the number of ones preceding it must be at least thirty-nine. That is, if there were thirty-eight or fewer then the addition of the twelve-vote bloc would yield a coalition with fewer than the fifty-one votes needed to make it a winning coalition. On the other hand, if the number of ones preceding the "12" were more than fifty, then the coalition would be winning before the "12" joined. Thus, the orderings that have "12" in the pivotal position are the ones with an initial sequence of ones of length $39, 40, \ldots, 50$. There are twelve numbers in this sequence. Hence, we can conclude the following:

$$\text{SSI(New England bloc)} = 12/89.$$

Thus, although the fraction of votes held by the New England bloc is only 12/100, the fraction of power (as measured by the Shapley–Shubik index) is 12/89.

The answer of 12/89 arrived at in the above example is easy to remember in terms of the parameters of the problem. That is, the numerator is just the size of the voting bloc, while the denominator is the number of distinct players when the bloc is considered to be a single player. The following theorem tells us that this is no coincidence.

THEOREM. *Suppose we have n players and that a single bloc of size b forms. Consider the resulting weighted voting body:*

$$[q : b, 1, 1, \ldots, 1]$$
$$\underbrace{}_{n - b \text{ of these}}$$

Assume $b - 1 \leq q - 1 \leq n - b$. Then the Shapley–Shubik index of the bloc is given by:

$$\text{SSI(bloc)} = \frac{b}{n - b + 1}.$$

PROOF. The argument is just a general version of what we did before. Notice first that $n - b + 1$ is just the number of distinct orderings. (Recall that we are not distinguishing between two orderings in which the ones have been rearranged.) Thus, we have $n - b$ ones and the number of places the b can be inserted is just one more than this. (in the example above, we had eighty-eight ones and eighty-nine places to insert the 12 bloc.)

The b bloc will be pivotal precisely when the initial sequence of ones is of length at least $q - b$ (since $q - b + b$ is just barely the quota q), but not more than $q - 1$ (or else the quota is achieved without b). Hence, the b bloc is pivotal when the initial sequence of ones is any of the following lengths:

$$q - 1, q - 2, \ldots, q - b.$$

Notice that since $q - 1 \leq n - b$ and $n - b$ is the number of ones available, we can construct all of these sequences—even the one with the initial segment requiring $q - 1$ ones. Clearly, there are exactly b numbers in the above list. Thus, the Shapley–Shubik index of the bloc of size b is given by:

$$\text{SSI}(b \text{ bloc}) = \frac{\text{number of orders in which } b \text{ is pivotal}}{\text{total number of distinct orderings}} = \frac{b}{n - b + 1}.$$

■ 9.7 CONCLUSIONS

We began this chapter by introducing two more power indices: the Johnston index and the Deegan–Packel index, and we calculated the power of the president (as well as the House and Senate) in the U.S. federal system according to these two power indices. The four power indices considered (two in Chapter 3 and two here) give strikingly different results. For example, the Banzhaf index suggests that the president has 4 percent of the power and the House holds roughly twice as much power as the Senate. The Johnston index, on the other hand, suggests that the president has 77 percent of the power with the remainder roughly divided 2 to 1 again, but with the Senate and House reversed from what we had with the Banzhaf index. So which of the two (or three or four) is more accurate, and how can we test this against actual experience with the federal government? Certainly, this is the right question to ask, but not of non–political scientists. Hence, we will content ourselves here with referring the reader to Brams, Affuso, and Kilgour (1989) and Packel (1981).

We also considered ordinal notions of power and introduced formal definitions intended to capture the intuitive idea of comparing the extent to which two voters are desired by coalitions that wish to become, or to remain, winning. The focus here was on a notion of when two voters have "incomparable" power (**Section 9.4**) or "comparable" power (**Section 9.5**). We concluded with a theorem on voting blocs.

EXERCISES

1. Consider the weighted voting system [8 : 5, 3, 1, 1, 1].
   (a) Calculate the Johnston index of each voter.
   (b) Calculate the Deegan–Packel index of each voter.
2. Consider a voting system for the six New England states where there are a total of seventeen votes and twelve or more are required for passage. Votes are distributed as follows:

   MA:4   ME:3   NH:2
   CT:4   RI:3   VT:1

   (a) Calculate the Johnston index of each voter.
   (b) Calculate the Deegan–Packel index of each voter.
3. Consider the mini-federal system with thirteen votes (used in **Section 3.6** wherein there are six House members, six senators, and the president. Passage requires half the House, half the Senate, and the president, or two-thirds of both houses.
   (a) Calculate the Johnston index of each voter.
   (b) Calculate the Deegan–Packel index of each voter.
4. Consider the minority veto system wherein there are eleven voters, three of whom are a designated minority, and passage requires a total of at least six of the eleven votes including at least two of the three minority votes.
   (a) Calculate the Johnston index of each voter.
   (b) Calculate the Deegan–Packel index of each voter.
5. Express $|T_{21}|$ in "n choose k" notation.
6. Express $|T_{23}|$ in "n choose k" notation.
7. Explain, in words, the "n choose k" expression of TJP(A senator).
8. Express the total Johnston power of a member of the House using the "n choose k" notation. (Mimic what we did for a senator in **Section 3.6**)

9.   Explain, in words, the expression giving the Deegan–Packel power of a House member and a senator.

10. Consider the yes–no voting system in which there are six voters: a, b, c, d, e, f. Suppose the winning coalitions are precisely the ones containing at least two of a, b, and c and at least two of d, e, and f.

   (a)   Show that a and b are equally desirable (as defined in **Section 9.4**)

   (b)   Show that the desirability of a and d is incomparable (using the definition in **Section 9.4**)

11. Suppose that x and y are voters in a yes–no voting system and that $x \approx y$. Suppose that Z' is a winning coalition to which both x and y belong. Assume that x's defection from Z' is critical. Prove that y's defection from Z' is also critical. (Hint: Assume, for contradiction, that y's defection from Z' is not critical. Consider the coalition Z arrived at by deleting x and y from Z'.)

12. Suppose that x and y are voters in a yes–no voting system and that $x \approx y$. Suppose that Z' is a coalition that contains x but not y. Let Z'' be the coalition resulting from replacing x by y in Z'.

   (a)   Prove that if Z' is winning, then Z'' is also winning.

   (b)   Prove that if Z' is losing, then Z'' is also losing.

   (Hint for both parts: Let Z be the result of deleting x from Z' and then argue by contradiction.)

13. Assume that x and y are voters in a *weighted* yes–no voting system.

   (a)   Assume that for some choice of weights and quota realizing the system, x and y have exactly the same weight. Prove that $x \approx y$.

   (b)   Assume that there are two choices of weights and quota realizing the yes–no voting system under consideration, one of which gives x more weight than y and one of which gives y more weight than x. Prove that $x \approx y$.

   [The converse of (a) and (b) is proved in Chapter 9, assuming the system is, in fact, weighted.]

14. In the U.S. federal system, let x be a member of the House and let y be a member of the Senate. Prove that x and y are incomparable.

15. In the U.S. federal system, let x be the vice president and let y be a member of the Senate. Prove that x and y are *not* incomparable. (One approach is to argue by contradiction.)

16. Using the results from Chapter 2, give an example of a yes–no voting system that is not weighted, but for which there are no incomparable voters.

17. Suppose x and y are voters in a yes–no voting system and suppose that their desirability is neither equal nor incomparable. Construct definitions (similar to what we did for incomparability) that formalize the notions that "x is more desirable than y" and "y is more desirable than x."

18. Prove that the relation of "equal desirability" as defined in **Section 9.5** is reflexive and symmetric. (This is part of the first proposition in **Section 9.5**)

19. Explain why $C - \{x\} \cup \{x\} = C$ if and only if x belongs to C.

20. Assume that we have two weightings, w and w', and two quotas, q and q', such that
    (i)   A coalition is winning if and only if $w'(Z) \geq q$.
    (ii) A coalition is winning if and only if $w'(Z) \geq q'$.
    (iii) $w(x) > w(y)$
    (iv) $w'(y) > w'(x)$.

    Assume that Z is an arbitrary coalition containing neither x nor y. Prove that $Z \cup \{x\}$ is winning if and only if $Z \cup \{y\}$ is winning.

21. Prove that, in the United States federal system, if x is a senator and y is the vice president, then $x > y$. (That is, prove that every senator is more desirable–according to the definition in **Section 9.5**—than is the vice president.)

22. Using results from earlier chapters, explain why a yes–no voting system is linear if and only if it is swap robust, and why every weighted voting system is linear. Is every linear voting system weighted?

23. Prove that, in a weighted voting system, we have $x > y$ if and only if x has strictly more weight than y in every weighting that realizes the system. (The reader will want to make use of earlier theorems to verify this.)

24. Consider the weighted voting body $[5 : 2, 2, 1, 1, 1, 1, 1]$. Calculate $SSI(x)$ where x is one of the people with two votes. (Hint: The theorem in **Section 9.6** does not directly apply here, but the underlying ideas of that theorem are all that is needed. That is, hold the "2" under consideration out, and ask how many distinct orderings of the remaining five 1s and one 2 there are. For each ordering, where is the insertion of the "2" under consideration pivotal?)

25. Use the theorem from **Section 9.6** (where it applies) to calculate the Shapley–Shubik index of the voting bloc in each of the following weighted voting bodies. If the theorem doesn't apply, say why.

   (a) [5 : 7, 1, 1, 1]
   (b) [8 : 3, 1, 1, 1, 1, 1, 1, 1, 1, 1, 1, 1, 1, 1, 1]
   (c) [6 : 3, 1, 1, 1, 1, 1, 1]

10

More Conflict

■ 10.1 INTRODUCTION

In this chapter, we continue our study of 2 × 2 ordinal games (and
variants thereof) with particular emphasis on game-theoretic models of international conflict. In **Section 10.2** we consider the joint
U.S.-Soviet policy of mutual assured destruction ("MAD") from the
1960s, 1970s, and 1980s. This treatment of deterrence tries to take
into account not only the actual preferences of each side, but also
each side's perception (perhaps better: fear) of the other's preferences.
In **Section 10.3** we return to the issue of deterrence and follow Brams
(1985a, 1985b) in considering a model of deterrence based on Chicken,
but with the choice of strategies being "probabilistic." This section
also introduces the ideas of cardinal utilities and expected value, thus
setting the stage for an introduction of 2 × 2 zero-sum games in
Section 10.4.

■ 10.2 MODELS OF DETERRENCE

In the 1960s, 1970s, and 1980s, both the United States and the Soviet
Union had the nuclear capability to destroy the other via a preemptive

first strike. So, in one very real sense, each was defenseless and at the complete mercy of the other. If this were the whole story, then rational action (suitably, although perhaps questionably, defined) would result in a race to the launch keys. But it's not the whole story. The point is, our ability to destroy each other did not include the capacity to destroy each other's ability to retaliate (although, of course, such retaliation has no effect on the destruction visited upon oneself), and it is precisely this ability to retaliate (and the mutual belief that such retaliation—even if "irrational"—is not inconceivable) that was intended to deter each side from initiating a first-strike attack. This mutual defense policy of deterrence is often referred to as "mutual assured destruction" and better known by its acronym "MAD."

Deterrence can be modeled as a 2×2 ordinal game with the United States and the Soviet Union being the two players, and each having the option to strike or not. Thus, the framework for this model is as in Figure 1 below.

The real question here is: What are the preference rankings for the four possible outcomes? The four possibilities in Figure 2 suggest themselves; we phrase each in terms of U.S. preferences, although we could equally well work with their analogues for Soviet preferences.

Notice that possibilities I and II are a pair in the sense of being the same except for having 1 and 2 switched. Similarly, possibilities III and IV are a pair (again with 1 and 2 switched). We can also regard I and III as a pair (having 3 and 4 switched) and II and IV as a pair

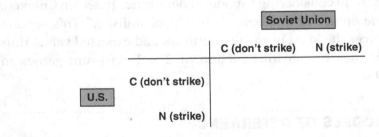

FIGURE 1

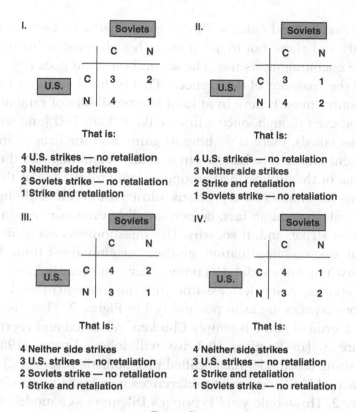

I.

		Soviets	
		C	N
U.S.	C	3	2
	N	4	1

That is:

4 U.S. strikes — no retaliation
3 Neither side strikes
2 Soviets strike — no retaliation
1 Strike and retaliation

II.

		Soviets	
		C	N
U.S.	C	3	1
	N	4	2

That is:

4 U.S. strikes — no retaliation
3 Neither side strikes
2 Strike and retaliation
1 Soviets strike — no retaliation

III.

		Soviets	
		C	N
U.S.	C	4	2
	N	3	1

That is:

4 Neither side strikes
3 U.S. strikes — no retaliation
2 Soviets strike — no retaliation
1 Strike and retaliation

IV.

		Soviets	
		C	N
U.S.	C	4	1
	N	3	2

That is:

4 Neither side strikes
3 U.S. strikes — no retaliation
2 Strike and retaliation
1 Soviets strike — no retaliation

FIGURE 2

(again with 3 and 4 switched). Thus, there are two questions giving rise to the four possibilities:

1. Would either country prefer an unanswered nuclear strike on the other to mutual coexistence?

2. If one country launched a first strike, would the other really respond even though this would do nothing to better its own situation?

We certainly don't have the answer to either question, and, indeed, the answers certainly need not have been constant throughout the 1960s, 1970s, and 1980s. Recollection of Khrushchev's promise to "bury you" or Reagan's characterization of the Soviet Union as the "evil empire" might suggest an affirmative answer to the first question (i.e., preference for a "successful" first strike), but one must also keep in mind that in the 1950s the United States had a first-strike nuclear capability and chose not to use it even though some influential people were encouraging its use. The second question goes right to the heart of the "paradox of deterrence." That is, for deterrence to work, each country must believe in at least the possibility of retaliation by the other, even though, once a first strike is launched (and so deterrence has failed), there is nothing to gain—and the lives of millions of innocent children to be lost—in actually retaliating. On the other hand, one of the authors had an opportunity to ask one of the chief U.S. arms negotiators for a previous administration if he thought the United States would, in fact, launch an all out nuclear response to a Soviet first strike, and, if so, why. The question was asked in a very informal, one-on-one situation, and we certainly don't think he was being anything but candid. His reply: "Yes; just because."

As a starting point, let's assume (for the moment) that U.S. and Soviet preferences are as in possibility I in Figure 2. Then the resulting 2 × 2 ordinal game is simply Chicken, relabeled and reproduced in Figure 3. [In **Section 10.3** we will follow Brams (1985b) in embellishing this model via so-called probabilistic responses.]

The actual U.S. and Soviet preferences may be as in possibility II in Figure 2. This would yield Prisoner's Dilemma as a model of deterrence, and, indeed, this has been proposed and well argued by Zagare (1987). Chicken, however, offers a more immediate explanation of the unstable situation of mutual cooperation of the 1980s. That is, we know that the (3,3)-outcome in Chicken is not a Nash equilibrium, and so each country has an incentive to unilaterally change strategy. It should also be noted that the Prisoner's Dilemma model predicts mutual first strikes, and this has not happened. There is, however, an

		Soviet Union	
		Don't strike	**Strike**
	Don't strike	(3, 3)	(2, 4)
U.S.			
	Strike	(4, 2)	(1, 1)

FIGURE 3

important issue that needs to be introduced at this point: If one is only talking about the actual U.S. and Soviet preferences, then a single 2×2 ordinal game suffices to display this information. If, however, one now wants to talk about strategies in the game, then the following question arises: Has each side correctly identified the other's preference ranking? The point is, the (true) preferences determine exactly which 2×2 ordinal game is actually being played. But strategies are based on the 2×2 ordinal game that each side *perceives* is being played, and this perception depends on a best guess by each as to the preferences of the other.

We think deterrence during, say, the 1970s provides an excellent example of misperception in the above sense. In particular, we think the actual preference rankings of both the United States and the Soviet Union at this time may well have been as in possibility IV in Figure 2 (with mutual cooperation most preferred). However, each perceived (perhaps better: feared) that the other had preference rankings as in possibility II in Figure 2 (with a successful first strike most preferred). This gives rise to the three 2×2 ordinal games in Figure 4 below.

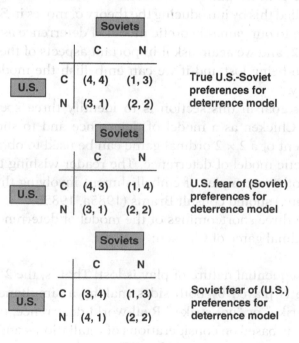

FIGURE 4

Looking, for example, at the middle game in Figure 4 we see an explanation for the tension of the times. If we again consider the status quo to be the (4, 3) mutual cooperation outcome, the United States will perceive the Soviets as wanting to unilaterally move from C to N, that is, (4, 3) is not a Nash equilibrium, and so we must retain a posture that suggests we prefer mutual noncooperation (i.e., retaliation) to one-sided nuclear annihilation.

Notice, however, that for deterrence to work, it is not necessary for each side to convince the other that it will definitely retaliate against a first strike. It is enough to convince the other side that one will probably (whatever that means) retaliate. In **Section 10.3**, we address deterrence in this context of uncertainty.

■ 10.3 A PROBABILISTIC MODEL OF DETERRENCE

While 2×2 ordinal games provide us with models that are easy to work with, they are often too simple in structure to capture much of what is going on with the real-world situation at hand. Our game-theoretic model of the Yom Kippur War in **Section 4.6** was one such example, and we handled this by introducing the theory of moves in **Section 4.7**. We now turn to our game-theoretic model of deterrence as Chicken in **Section 10.2**, and we again ask if important aspects of the real-world situation are being lost and if we can embellish the model to better reflect reality.

Thus, our goal in this section is to identify three specific short-comings of Chicken as a model of deterrence and to show how an embellishment of a 2×2 ordinal game can be used to obtain a better game-theoretic model of deterrence. The reader wishing to see more on this (or other examples of embellishment involving threat power, deception, etc.) should consult Brams (1985a, 1985b).

There are three shortcomings of the model of deterrence based on the 2×2 ordinal game of Chicken:

1.   The sequential nature of play is lost. That is, the 2×2 ordinal game is played by both sides making a simultaneous choice to "strike" or "not strike." Real-world deterrence, on the other hand, is based on considerations of retaliation—and retaliation

is certainly a concept that only arises in the context of sequential events.

2. Ordinal preferences contain less information than cardinal utilities. Recall that when we first introduced 2 × 2 ordinal games in **Section 4.2**, we made a point of emphasizing that a preference ranking of 4 for an outcome should not be construed as indicating that it is considered twice as good as the outcome ranked 2. That is, the numbers 4, 3, 2, and 1 were ordinal preferences—numbers indicating the order of preference of outcomes without saying anything about the absolute "worth" of the outcome. If we replace the 4, 3, 2, 1 with numbers that are supposed to measure the absolute worth of the outcomes (in units that are left unspecified), then we're in the realm of cardinal utilities. We can, in fact, always assume that the units have been chosen so that the cardinal utility of the most preferred outcome is one (unit) and that of the least preferred outcome is zero (units). In between, then, we may have cardinal utilities like $\frac{1}{3}$ and $\frac{2}{3}$, or .01 and .02, or .98 and .99, or even .01 and .99. Interpreting the last one, for example, we see that the two least preferred outcomes are considered about the same (0 utility versus .01 utility) while the most preferred pair is considered much more desirable than is the least preferred pair.

3. Aspects of uncertainty are lost. A single 2 × 2 ordinal game presents a situation of complete information, at least as it applies to the preferences of one's opponent. However—and this is what we tried to illustrate with the models of deterrence in Figure 4—in real life one typically can only make educated guesses as to the preferences of one's opponent. Phrases like "Chances are three to one that he would prefer not to strike even if we did" are not uncommon in this context.

Continuing with this last point, let's convert the phrase "chances are three to one . . ." to a phrase involving probabilities. "Three to one" really means "three out of four." So when we say, "Chances are three to one something will happen," we really mean, "In the long run (i.e., if this situation were re-created many many times), we'd expect that the something referred to would happen in (approximately) three-fourths of the situations (briefly, three-fourths of the time)." Probabilities simply provide another terminology for expressing the same thing. To say, "The probability of something happening is three-fourths," is the same as saying, "The chances of this thing happening are three to one." Both are meant to convey the idea that, if the situation were re-created many many times, then the thing being referred to would take place in about 75 percent of the situations.

The final thing we have to do before presenting the models we wish to consider is to give at least an informal treatment of something called "an expected value calculation." To illustrate this, let's take a simple example. Suppose that we plan to roll a single die, and, if it comes up 1 or 2, to give you ten dollars, and if it comes up anything else, to give you thirty dollars. Suppose we repeat this over and over again. Then it makes sense to ask what you'd expect the average amount of money changing hands to be for this procedure. To see what the answer to this is, notice first that a roll of 1 or 2 will occur about one-third of the time, while a roll of something else will occur about two-thirds of the time. Thus, in the long run, you can expect to get ten dollars about one-third of the time, and thirty dollars two-thirds of the time. Hence, the average per roll that you will receive is given by the following:

$$(\$10) \times [\text{fraction of time we roll a one or two}] +$$

$$(\$30) \times [\text{fraction of time we roll something else}]$$

which equals

$$(\$10) \times \frac{1}{3} + (\$30) \times \frac{2}{3} = \frac{10}{3} + \frac{60}{3} = \frac{70}{3} = 23\frac{1}{3} \text{ dollars.}$$

Instead of the "average you'd receive in the long run," probabilistic terminology describes this as "an expected value of $23\frac{1}{3}$ dollars." Notice also that the "$\frac{1}{3}$" and "$\frac{2}{3}$" occurring above is a special case of the general observation that if $0 \le p \le 1$ and p is the probability of something happening, then $1 - p$ is the probability of it not happening [since $p + (1 - p)$ equals one].

Now, let's see how we can change the model of deterrence from one based on the 2×2 ordinal game of Chicken to one based on a new (and slightly more complicated) game that will deal with the three shortcomings listed above.

The "game board" is similar to that used for a 2×2 ordinal game in that we have the same two players (Row and Column) and the same two choices of strategy (C or N). Outcomes, however, are now labeled by each player not just with preference rankings 1, 2, 3, and 4, but with numbers giving the cardinal utility of that outcome for that player. We assume that the units with which utility is being measured are chosen so that a least preferred outcome has utility zero and a most preferred outcome has utility one. We'll use "r's" for Row's utilities and "c's" for Column's. Thus, Row has utilities 0, r_2, r_3, and 1, while Column has utilities 0, c_2, c_3, and 1. As an example—and this is the one based on Chicken that we will be using—we might have Figure 5.

Be careful not to confuse "utility one," which is best possible, with "ordinal preference ranking one," which is worst possible.

Given the game board, and we'll continue here with the one in Figure 5 for example, the game is played as follows:

1. Row goes first and chooses either C or N.

2. If Row chooses C, then so does Column and the outcome is (r_3, c_3).

3. If Row chooses N, then Column does the following:
   (a) With probability p, Column chooses C and the outcome is $(1, c_2)$. Intuitively, Column has chosen to "wimp out" and not retaliate, and thus get his next to worst outcome of c_2 units utility while allowing Row to get his maximum utility of one.

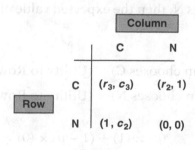

FIGURE 5

(b) With probability $1 - p$, Column retaliates by choosing N also. This yields an outcome of $(0, 0)$.

This describes the game board and the play of the game. The interesting question is: What does rationality dictate Row's behavior should be? (Notice that Column's behavior is predetermined.) Intuition suggests the following.

1. A large value of p (that is, a value of p close to 1) suggests that Column has very little resolve, and so Row should tend toward choosing the aggressive strategy of N. That is, in this situation it might be reasonable for Row to sacrifice the sure r_3 utility of the CC outcome he can guarantee by initially cooperating, since he stands a good chance of getting utility 1 from the NC outcome that will result if Column chooses not to retaliate. The risk is that Row gets zero utility if Column does retaliate.

2. A large value of r_3 (that is, a utility value close to the top utility value of 1) suggests that Row should play it safe and choose C, thus getting the guaranteed outcome of CC and the utility r_3 (which we're assuming is only slightly less than the top utility of 1 that Row might get by risking a utility outcome of zero).

If we do the expected value calculation, this intuition is borne out. That is, the question of whether Row should choose C or N depends on whether r_3 is bigger than p or not. Row's expected value in terms of utility is calculated as follows.

1. If Row chooses C, then the utility for Row is r_3.

2. If Row chooses N, then the expected value of the utility for Row is given by:

(Probability Column chooses C) $\times$ (Utility to Row of this choice) $+$

(Probability Column chooses N) $\times$ (Utility to Row of this choice) $=$

$$(p) \times (1) + (1 - p) \times (0) =$$
$$p + 0 = p.$$

Thus, p turns out to be the expected value of the utility that Row achieves by choosing the aggressive strategy of N. Hence, rational action dictates that Row should

1. choose C if $r_3 > p$ (and capitalize on the high utility—relative to the probability that Column will not retaliate—of the CC outcome),

2. choose N if $p > r_3$ (and capitalize on the high probability—relative to the utility of the CC outcome—that Column will not retaliate).

A discussion of the above game in the context of models of deterrence is asked for in Exercise 6 at the end of the chapter.

■ 10.4 TWO-PERSON ZERO-SUM GAMES

In this section, we consider 2×2 games that are called "zero-sum games." These games differ in three fundamental ways from the 2×2 ordinal games like Prisoner's Dilemma that we dealt with earlier:

1. Outcomes will be expressed in terms of cardinal utilities that measure the value of each outcome (perhaps in dollars, perhaps not) to each player. Hence, an outcome with utility 4 for Row is, in fact, twice as good for Row as an outcome with utility 2. We also allow outcomes to have negative utility for either player: an outcome with utility -4 for Row is twice as bad for Row as one with utility -2.

2. The game is *zero sum*: the utility of any outcome to Row plus the utility of that same outcome to Column is zero. Thus, Column's utilities are obtained from Row's utilities by simply deleting the minus sign from the negative numbers and adding a minus sign to the positive numbers.

3. Each player will have two "pure strategy" options (which we shall continue to call C and N except when modeling real-world situations). But in addition to this, each player will have the

option of choosing some number (e.g., a fraction) p strictly between 0 and 1 as a strategy. These are called *mixed strategies*. We interpret a choice of p as yielding a play wherein that player employs a random device to determine whether he or she will play C or N, and that this is done so that the probability that C will be chosen is p. For example, if $p = \frac{1}{4}$, the player might flip two coins and play C if and only if both come up tails. Or, if $p = \frac{1}{10}$, he or she may start a digital stopwatch, stop it a few seconds later, and then play C if and only the number of seconds elapsed is of the form 3.06 or 3.16 or ... or 3.96. (There will be, on average, a "6" in the hundredths place one-tenth of the time.)

As a particular example of such a game, consider the following situation involving military intelligence. Suppose Column is a country with two military installations, A and B, and resources sufficient to defend only one of the two. Suppose Column is engaged in a war with Row, and that Row has the resources to attack either A or B, but not both. Suppose that installation A is three times more valuable to both than installation B. Our starting point is to model this situation with a 2 × 2 game where the payoffs are not ordinal rankings but cardinal utilities. We assume that an attack of a defended position results in neither a gain nor a loss for either player.

		Column	
		Defend A	**Defend B**
	Attack A	(0, 0)	(3, –3)
Row			
	Attack B	(1, –1)	(0, 0)

The game is certainly zero sum. In fact, it is clear that we really don't need the ordered pair notation; if we simply give the payoffs to Row, then we can obtain the payoffs to Column immediately. Thus, the standard presentation of the above game would be:

	Column	
	Defend A	Defend B
Attack A	0	3
Attack B	1	0

Notice that a positive entry represents a gain for Row and a loss for Column. A negative entry (of which there are none in this game) represents a loss for Row and a gain for Column. Hence, this game favors Row.

The question is, What is Column's best strategy?

Wrong answer #1: Column says: "Row expects me to defend A, so he will attack B. But Row knows that I will reason this way, and so—assuming I therefore will defend B—he will attack A. But Row also knows that 1 will reason this way, and so he will ..."

Wrong answer #2: Flip a coin. Heads—defend A. Tails—defend B.

We shall see that, in some sense, wrong answer #1 asks too much of mathematical analysis while wrong answer #2 asks too little of mathematical analysis. That is, #1 suggests that some kind of game-tree analysis will, in fact, reveal one of the pure strategies to be Column's best choice. This turns out not to be the case. On the other hand, #2 suggests that if neither pure strategy is the best choice, then one simply throws up one's hands and randomizes (with $p = \frac{1}{2}$). This also turns out to be false. The truth of the matter, for this particular game, is given in the following proposition:

PROPOSITION. *If Column chooses $p = \frac{3}{4}$ (that is, if Column chooses to defend installation A with probability $\frac{3}{4}$, and to defend installation B with probability $\frac{1}{4}$), then he will obtain an expected value no worse than $-\frac{3}{4}$, regardless of what pure or mixed strategy Row chooses to employ. For any choice other than $p = \frac{3}{4}$ by Column, there is a strategy for Row (in fact, a pure strategy) that leaves Column with a worse expected value than $-\frac{3}{4}$.*

The choice of $p = \frac{3}{4}$ in the proposition is called Column's *minimax strategy*: it minimizes the maximum amount Row can expect to get in the long run (and thus maximizes the minimum amount Column can expect to get in the long run).

PROOF. The proof consists of the following five claims. For claims 1 to 3, we are supposing that Column chooses $p = \frac{3}{4}$.

CLAIM 1. If Row chooses the pure strategy "Attack A," then the expected value of the payoff to Column is $-\frac{3}{4}$.

PROOF. Picture hundreds and hundreds of plays of this game with Row always choosing to attack A, and Column defending A three-fourths of the time. Then three-fourths of the time (on average) the payoff to both Row and Column is zero, since Row's attack of A is met by Column's defense of A. In the other quarter of the cases, however, Row is gaining a utility of 3 (and thus Column is "gaining" a utility of −3) because Row is attacking A while Column is defending B. Thus Row's expected value is:

$$\frac{1}{4} \times 3 = \frac{3}{4},$$

and so Column's expected value must be $-\frac{3}{4}$, as claimed.

CLAIM 2. If Row chooses the pure strategy "Attack B," then the expected value of the payoff to Column is $-\frac{3}{4}$.

PROOF. Arguing as in Claim 1, we see that Row gains a utility of one by attacking B in the (roughly) three-fourths of the plays where Column defends A. Row gains nothing in the other quarter wherein he is attacking B while Column is defending B. Thus, Row's expected value is:

$$\frac{3}{4} \times 1 = \frac{3}{4},$$

and so Column's expected value must again be $-\frac{3}{4}$, as claimed.

CLAIM 3. If Row chooses a mixed strategy of q, then the expected value of the payoff to Column is $-\frac{3}{4}$.

PROOF. As before, let's picture many many plays of the game, but now with Row sometimes attacking A and sometimes attacking B. This means that some fraction of the time we are in the situation dealt with in claim 1 (in fact, this fraction is q) and some fraction of the time $(1-q)$ in the situation dealt with in claim 2. But now it is easy to calculate the expected value for Column (since the results in claim 1 and claim 2 were the same). It is simply

$$[q \times (-\frac{3}{4})] + [(1-q) \times (-\frac{3}{4})] = -\frac{3}{4},$$

as claimed.

CLAIM 4. If x is some small positive number, and Column chooses the mixed strategy $\frac{3}{4} + x$, then Column's expected value will be strictly worse than $-\frac{3}{4}$ if Row chooses the pure strategy of attacking B.

PROOF. This is Exercise 7 at the end of the chapter.

CLAIM 5. If x is some small positive number, and Column chooses the mixed strategy $\frac{3}{4} - x$, then Column's expected value will be strictly worse than $-\frac{3}{4}$ if Row chooses the pure strategy of attacking A.

PROOF. This is Exercise 8 at the end of the chapter.

This completes the proof of the proposition.

In general, it is quite easy to prescribe "optimal play" in 2×2 zero-sum games. The following theorem pretty much tells the whole story.

THEOREM. *Suppose we have a 2×2 zero-sum game:*

	C	N
C	a	b
N	c	d

Assume neither player has a dominant strategy. (We assume that if either player has a dominant strategy, then the reader can prescribe optimal play for both Row and Column.) Suppose a is the largest entry in the matrix and d is the second-largest entry in the matrix.

(If there are no dominant strategies, this can always be achieved by interchanging the columns and/or interchanging the rows.) Then:

1. The minimax strategy for Column is to play column 1 with probability p, where

$$p = \frac{d - b}{(a - b) + (d - c)}.$$

2. The minimax strategy for Row is to play row 1 with probability q, where

$$q = \frac{d - c}{(a - b) + (d - c)}.$$

3. The expected payoff to Row is

$$V = \frac{ad - bc}{(a - b) + (d - c)}.$$

We omit the proof of this theorem, although the exercises contain several applications of it. The V occurring in part 3 of the theorem is called the *value* of the game. We should also point out that by restricting our discussion to the 2 by 2 case, we have missed an opportunity to discuss so-called saddle points of games. This is somewhat rectified in the exercises.

We close our discussion of 2×2 zero-sum games by stating two remarkable theorems, both of which apply to the case where both players have several strategies (instead of just two) available to them. The first is the celebrated minimax theorem of John von Neumann (1928). The second is a beautiful result due to Julia Robinson (1951).

THEOREM. *(John von Neumann, 1928). Suppose G is a two-person zero-sum game with finitely many strategies available to each player. Then there exists a number V, called the value of the game, such that*

1. *Row has a strategy (pure or mixed) that will give him an expected value of at least V regardless of what (pure or mixed) strategy Column uses, and*

2. *Column has a strategy (pure or mixed) that will give him an expected value of at least −V regardless of what (pure or mixed) strategy Row uses.*

THEOREM. *(Julia Robinson, 1951). Suppose G is a game as in the minimax theorem, above. Suppose G is played over and over again with each player choosing the pure strategy that maximizes his expected payoff against the "accumulated mixed strategy of his opponent up to then." [By this, Robinson (1951) means that each player is assuming that the probability that a given pure strategy will be used against him at stage n + 1 is equal to the fraction of times that strategy was chosen in the first n stages.] Then these strategies converge to the optimal mixed strategies guaranteed to exist in the minimax theorem.*

Much more on zero-sum games can be found in Straffin (1993).

■ 10.5 CONCLUSIONS

Section 10.2 and **10.3** both involved models of deterrence. The first section gave a somewhat standard treatment based on a 2 × 2 game, but with an extended discussion of the problems caused if each side misperceives the preferences of the other. **Section 10.3**, on the other hand, contained quite a different treatment, and it set the stage for a discussion of 2 × 2 zero-sum games in **Section 10.4**.

EXERCISES

1. On one page, write your thoughts on the two questions that follow Figure 2 in **Section 10.2**.
2. Discuss dominant strategies and Nash equilibria for the three 2 × 2 ordinal games in Figure 4 in **Section 10.2**.
3. Consider the hypothetical situation where people fill out their federal income tax forms based on expected value calculations as opposed to the patriotic and moral considerations that are so dominant in real life. We used to have the opportunity to claim a $100 tax exemption (i.e., an outright saving of $100) for contributions to political candidates. Suppose 1 didn't contribute, but 1 also knew that only

7 percent of those who falsify their returns in this way were caught by the I. R. S. What is the smallest fine that should have dissuaded me from falsifying my return in this way?

4. Generalize the result in the previous exercise so as,to arrive at a formula that gives the minimum penalty necessary to discourage (in terms of expected value) the breaking of a law. The formula should involve the probability of being caught and the gain achieved by breaking the law.

5. Redo the probabilistic model of Chicken assuming that the utilities are r_1, r_2, r_3, r_4 and c_1, c_2, c_3, c_4.

6. Do an analysis of "probabilistic Prisoner's Dilemma" that mimics what we did for Chicken.

7. In the attack-defend game from **Section 10.4**, prove that if x is some small positive number, and Column uses the mixed strategy $\frac{3}{4} + x$, then Column's expected value will be strictly worse than negative $\frac{3}{4}$ if Row chooses the pure strategy of attacking B. (This is Claim 4 in the proposition in **Section 10.4**.)

8. In the attack-defend game in **Section 10.4**, prove that if x is some small positive number, and Column uses the mixed strategy $\frac{3}{4} - x$, then Column's expected value will be strictly worse than negative $\frac{3}{4}$ if Row chooses the pure strategy of attacking A. (This is claim 5 in the proposition in **Section 10.4**.)

9. Suppose we have a 2×2 zero-sum game:

	C	N
C	a	b
N	c	d

Assume that a is the largest entry and assume that d is not the second-largest entry. Show that either Row or Column has a dominant strategy. (Assume, for simplicity that the numbers a, b, c, and d are distinct.)

10. Consider the following 2×2 zero-sum game:

	C	N
C	a	b
N	c	d

Assume that a is the largest entry and that d is the second-largest entry. Notice that if Column plays C with probability p, then Column receives

$$p(-a) + (1-p)(-b) \quad \text{if Row plays C,}$$

and

$$p(-c) + (1-p)(-d) \quad \text{if Row plays N.}$$

(a) Set these two expressions equal to each other and solve for p.

(b) Explain the relevance of this to the theorem for 2×2 zero-sum games in **Section 10.4**.

(c) Repeat (a) and (b) for Row.

11. Find the optimal strategies and value for the following games. (If there is a dominant strategy, then the value of the game is defined to be the payoff for Row when the dominant strategy is used by whichever player has it, and the other player responds with his or her best choice of pure strategy.)

	C	N		C	N		C	N
C	2	1	C	-1	-2	C	4	1
N	-1	3	N	3	4	N	-2	2

12. An outcome in a two-person zero-sum game, even allowing for the possibility of more than two strategy choices for each player, is called a *saddle point* if it is simultaneously the smallest (or tied for such) entry in its row and the largest (or tied for such) entry in its column. Prove that an outcome is a saddle point if and only if it is a Nash equilibrium (in pure strategies).

13. Find the saddle points in the following games:

	C	N		C	N	V		C	N	V
C	-1	-2	C	1	-1	2	C	3	1	4
N	3	4	N	3	1	-2	N	1	0	-2
			V	1	0	2	V	2	1	3

14. In the book *Superior Beings*, the author, Steven J. Brams (1983) of New York University's Department of Politics, asks the following question: If God is omniscient ("all knowing"), would we be able to determine if He had this power via our interactions with Him? Brams methodology consists of
    (a) Rigorously defining omniscience game-theoretically,
    (b) Modeling our relationship with God game-theoretically,
    (c) Analyzing the game in 2 ways: assuming God is not omniscient and then assuming God is omniscient,
    (d) Concluding that if the outcomes are different then we can detect the power if He has it.

    The fundamental game that Brams uses to model our relationship with God is the so-called Revelation Game:

		People	
		Believe in God	**Don't Believe**
God	**Reveal Himself**	(3, 4)	(1, 1)
	Don't Reveal Himself	(4, 2)	(2, 3)

We take these preferences as given.
    (a) Analyze the game. That is, does either God or People have a dominant strategy and, if so and the other side knows it, what will the outcome be?
    (b) Assume that God is omniscient and that this means, game-theoretically, that God knows which strategy People will choose. This is equivalent to saying that people move first and then God will respond with His move (and the game ends). Analyze this version of the game.
    (c) Can we determine if God is omniscient by this interaction?
15. In his book *Biblical Games*, Brams (1980) considers the story of Samson—a "ferocious warrior of inhuman strength"—and his wife Delilah who was paid by the Philistines to find out the secret of Samson's strength. Brams models the situation as follows:

	Samson	
	Tell	Don't Tell
Don't Nag	(4, 2)	(2, 4)
Nag	(3, 3)	(1, 1)

Delilah

We take these preferences as given. Analyze this situation in two different ways:

(a) As a 2 × 2 game played as usual, what does this predict will happen?

(b) By the theory of moves starting at (2, 4) with Delilah going first. Assume that a move from (4, 2) to (2, 4) by Samson is impossible, since he can't take back "telling." What does this predict will happen?

More Fairness

....

■ 11.1 INTRODUCTION

In this chapter, we continue our study of fair division. We start with a closer look at the adjusted winner procedure, introduced in **Section 5.6**, that guarantees an efficient, equitable, and envy-free allocation of goods for two people. In **Section 11.2**, we will prove the efficiency of the procedure, and in **Section 11.3**, we will see that, typically, honesty is the best strategy for the procedure.

Beginning in **Section 11.4**, we study cake-cutting procedures for three or more people. Just as the addition of a third alternative considerably complicates the search for a perfect social choice procedure, we will see that the addition of a third party also complicates the search for the best fair division method. We present in **Section 11.4** a procedure that guarantees each of three parties a proportional share of cake. In **Section 11.5**, we consider two procedures which each guarantee an envy-free portion of cake. In **Section 11.6**, we present an envy-free procedure for four parties. None of the methods we consider here are efficient nor equitable, however, and in **Section 11.7**, we will see that our search for an efficient, envy-free, and equitable procedure for three or more parties will fare no better than our search for a perfect social choice procedure or apportionment method.

■ 11.2 EFFICIENCY IN ADJUSTED WINNER

In **Section 5.6**, we saw that adjusted winner gives an allocation of indivisible items for two people that is equitable and envy-free. We now give the proof that adjusted winner guarantees an efficient allocation as well, probably one of the most remarkable properties of the procedure.

Recall that in the first stage of adjusted winner, every item is first given to the person who valued it most. Items are then transferred from the initial winner to the other party until both have an equal number of points. The proof of efficiency hinges on the order in which the items are transferred: the transfer begins with the item with the smallest ratio of points given by the initial winner to points given by the other party. In this way, we minimize the effective cost to the initial winner for all points transferred to the other party. Intuitively, the adjusted winner procedure is efficient because the initial stage of the procedure is efficient, and then efficiency is not affected during the equitability adjustment.

We will prove the efficiency of adjusted winner with the following three lemmas. Call the two parties Annie and Ben. We will use the notation $G_1, \ldots, G_n$ to denote the items to be divided between Annie and Ben. Let a_i and b_i denote the fractions of item G_i that Annie and Ben, respectively, receive in a given allocation of items. Let A_i and B_i denote the points awarded to item G_i by Annie and Ben, respectively.

LEMMA 1. *Suppose that we have an allocation of the items in which*

  (i) *Annie values item G_i at least as much as Ben does*

  (ii) *Ben values item G_j at least as much as Annie does*

Suppose that Annie trades her portion of G_i for Ben's portion of G_j. If this trade is strictly better for one player, then it is strictly worse for the other.

PROOF. Since Annie values item G_i at least as much as Ben does, then we know that $A_i \geq B_i$. Similarly, since Ben values item G_j at least as much as Annie does, then we know that $B_j \geq A_j$. We can ignore all items except G_i and G_j since they are not involved in the trade. During the trade, Annie gives away a total of $a_j A_i$ points, and gains a total of $b_j A_j$ points. If the trade is strictly better for Annie, then

(1) $$b_j A_j > a_i A_i.$$

Notice that

Ben's points after trade - Ben's points

$$\text{before trade} = a_i B_i - b_j B_j$$
$$\leq a_i A_i - b_j B_j \text{ since } B_i \leq A_i$$
$$\leq a_i A_i - b_j A_j \text{ since } B_j \geq A_j$$
$$< 0 \text{ by (1)},$$

so Ben is strictly worse off after the trade. Similarly, if the trade is strictly better for Ben, then it is strictly worse for Annie.

LEMMA 2. *Suppose that we have an allocation of the items in which $\frac{A_j}{B_j} \leq \frac{A_i}{B_i}$. If Annie trades her portion of G_i for Ben's portion of G_j, and this trade is strictly better for one player, then the trade is strictly worse for the other.*

PROOF. As in the proof of Lemma 1, if the trade is better for Annie, then $b_j A_j > a_i A_i$. Since $A_j/B_j \leq A_i/B_i$, then $A_j B_i \leq A_i B_j$. Then

Ben's points after trade - Ben's

$$\text{points before trade} = a_i B_i - b_j B_j$$
$$< B_i \left(\frac{b_j A_j}{A_i} \right) - b_j B_j \text{ since } b_j A_j > a_i A_i$$
$$= b_j \left(\frac{B_i A_j - B_j A_i}{A_i} \right)$$
$$< 0 \text{ since } A_j B_i \geq A_i B_j,$$

so Ben is strictly worse off after the trade. If the trade strictly benefits Ben, however, then it follows that $a_i B_i > b_j B_j$. Then

Annie's points after trade - Annie's

$$\text{points before trade} = b_j A_j - a_i A_i$$

$$< b_j A_j - A_i \left(\frac{b_j B_j}{B_i} \right) \text{ since } a_i B_i > b_j B_j$$

$$= b_j \left(\frac{A_j B_i - A_i B_j}{B_i} \right)$$

$$\leq 0 \text{ since } A_j B_i \leq A_i B_j,$$

so Annie is strictly worse off after the trade.

LEMMA 3. *If a given allocation is not efficient, there there exist goods G_i and G_j and some portions thereof such that if Annie exchanges her fraction a_i of G_i for Ben's fraction b_j of G_j, the resulting trade yields an allocation that is at least as good for both players and strictly better for at least one of the players.*

The proof of this lemma requires the assumption of *weak additivity* of preferences: if A and B are disjoints sets of goods, and Annie values A at least as much as some set X of goods and B at least as much as some set Y of goods, then she must value $A \cup B$ at least as much as $X \cup Y$.

PROOF. Since the given allocation is not efficient, there is an alternative allocation that is at least as good for both Annie and Ben and strictly better for at least one of the two, say Annie. So there exist disjoint sets S and T of goods belonging to Annie and Ben, respectively, such that an exchange of S for T makes Annie better off without hurting Ben. We just need to show that S and T can each be taken to be (possibly a fraction of) a single item. Write $S = S_1 \cup \ldots \cup S_n$, where the S_i's are pairwise disjoint, and each is a fraction of a single item. Ben can now break up T into a disjoint union $T = T_1 \cup \ldots \cup T_n$ (not necessarily subsets of a single item) such that an exchange of S_i for T_i yields an allocation that is no worse for him than the current allocation. It suffices for Ben to choose each T_i such that the value to Ben of T_i is no more than the value to Ben of S_i; this must be possible by weak additivity of preferences.

We claim that there exists an i such that Annie prefers the allocation obtained by exchanging S_i for T_i to the existing allocation. If such an i did not exist, then the existing allocation is at least as good for Annie as the one obtained by exchanging S_i for T_i for all i. By additivity of preferences again, then the existing allocation is at least as good for Annie as the one obtained by exchanging $S_1 \cup \ldots \cup S_n = S$ for $T_1 \cup \ldots \cup T_n = T$, contrary to assumption. Re-labeling if necessary, suppose that Annie prefers the allocation obtained by exchanging S_1 for T_1 to the existing allocation. Now S_1 consists of some portion of a single good, but T_1 may consist of portions of several goods. Write T_1 as a disjoint union of sets $T_{11} \cup \ldots \cup T_{1m}$ and S_1 as a disjoint union of sets $S_{11} \cup \ldots \cup S_{1m}$ such that each T_{1j} is some portion of a single good and Annie is better off with the allocation obtained by exchanging S_{1j} for T_{1j} than with the existing allocation. This can be done for the same reason as above.

Now also by the same reasoning as above, there must exist a j such that the allocation obtained by exchanging S_{1j} for T_{1j} is at least as good for Ben as the existing allocation. Otherwise, the existing allocation is better for Ben than the one obtained by exchanging S_{1j} for T_{1j} for all j. It follows by additivity of preferences that the existing allocation is better for Ben than the one obtained by exchanging S_1 for T_1, which is a contradiction. Thus, we have found subsets S_{1j} and T_{1j} each consisting of a portion of a single item for which a trade of S_{1j} for T_{1j} yields an allocation that is strictly better for Annie and no worse for Ben than the existing allocation. This completes the proof of the lemma.

Proof of Efficiency

We are now ready to prove that adjusted winner always yields an allocation that is efficient. Suppose otherwise. By Lemma 3, there exist goods G_i and G_j and portions thereof such that if Annie exchanges her fraction a_i of G_i for Ben's fraction b_j of G_j, the resulting trade yields an allocation that is at least as good for both and strictly better for at least one. Suppose that Annie was the initial winner after the first step of the adjusted winner procedure. Since Annie still has at least a_i of item G_i after any necessary transfers, then Annie must value item G_i at least as much as Ben does, so $A_i \geq B_i$. Now if Ben values item G_j at least as much as Annie does, then Lemma 1 implies that the trade will not benefit both parties as we are assuming. So it must be the case that Ben values item G_i less than Annie does, that is, $B_j < A_j$. But Ben has part of G_j, so he must have received that during the transfer stage of the adjusted winner procedure. Since only one item is split among the parties during adjusted winner, then G_j must be that item, so Annie must have all of item G_i. Since item G_i was not involved in the transfer stage, it follows that the ratio of points for item G_i in the adjusted winner procedure is at least as big as the ratio of points for item G_j. Thus $A_i/B_i \geq A_j/B_j$. By Lemma 2, this contradicts our assumption that the trade does not hurt either party. This completes the proof that the adjusted winner procedure is efficient.

■ 11.3 ADJUSTED WINNER AND MANIPULABILITY

Dispute resolution is often a stressful time for the parties involved. Determining point totals is itself not an easy task. The situation is still more stressful if the parties involved need to worry about strategies as well, especially in the case of a divorce where each party has in depth knowledge of the other's likes and dislikes. It is natural to wonder whether this knowledge would enable one party to manipulate the system, and achieve a better outcome by submitting disingenous point allocations. Another advantage of the adjusted winner procedure is that unless knowledge of the other's party's valuations is strictly one-sided, then honesty is the best policy.

Suppose that Annie and Ben are getting a divorce, and currently share the following items: a townhouse in Central Square, season passes to the Red Sox, and a painting by Klee. They value the items as follows:

Annie	Item	Ben
50	Townhouse	30
20	Red Sox Tickets	50
30	Klee painting	20
100	Total	100

Applying the adjusted winner procedure, we see that Annie is initially awarded the townhouse and the painting, while Ben gets the

Red Sox tickets. Annie currently has 80 points, while Ben has 50, so Annie is the initial winner. The ratio of points for the townhouse is 5/3, while the ratio for the painting is 3/2, so the painting needs to be divided. Solving for x in the following equation gives the fraction of the painting that Annie keeps:

$$50 + 30x = 50 + 20(1 - x) = 70 - 20x$$
$$50x = 20$$
$$x = 2/5$$

Annie ends up with the townhouse and 2/5 of the painting (Annie and Ben decide that she will buy out his share of the painting), and Ben gets the Red Sox tickets and 3/5 of the painting—each with a total of 62 points. Now Annie has known Ben for ten years, and knows how much his Red Sox tickets mean to him. She is confident that she can estimate Ben's point allocations fairly well, and decides to submit the following false valuations, rather than her true preferences given above.

Annie's Fake Valuations	Item	Ben's Sincere Valuations
32	Townhouse	30
48	Red Sox Tickets	50
20	Klee painting	20
100	Total	100

Intuitively, Annie might do better under this scenario. By indicating that she values the townhouse only slightly more than Ben, she hopes to win the townhouse but at a lower cost, thereby winning a higher percentage of the painting as well. This time, in the first step of the process, Annie still gets the townhouse and the painting, and Ben gets the Red Sox tickets. Annie has 52 points (according to her false point allocations), and Ben has 50. Solving for x gives the fraction of the painting that Annie keeps:

$$32 + 20x = 50 + 20(1 - x) = 70 - 20x$$
$$40x = 38$$
$$x = 19/20$$

Under these false pretenses, then, Annie keeps 19/20 of the painting, only needing to pay Ben 1/20, 5%, of the painting's appraised value, rather than the 60% of the value that she would have paid had she given her true point allocations. Given the value of a Klee painting, Annie is saving a significant amount of money! In terms of points, Ben ends up with $50 + 1/20(20) = 51$ points, significantly less than the 62 points he had before. Annie appears to also have 51 points, but according to her true valuations, she really gets $50 + 19/20(30) = 78.5$ points.

It is definitely to Annie's advantage to submit false point allocations in the scenario above—but we were assuming that Ben would submit his honest valuations. If Annie were really able to predict Ben's point scheme so well, then it is a fair assumption that Ben would also be able to guess how Annie valued the items. With this kind of knowledge on both sides, it becomes much riskier to submit false preferences; while it may be to someone's advantage to be dishonest (Annie might still get lucky if Ben chooses to submit his true point allocations even with knowledge of Annie's preferences), this strategy can also backfire, resulting in an outcome that is worse than the honest outcome. For example, if Ben thinks that Annie will be honest, he may submit the following point allocations:

Annie's Sincere Valuations	Item	Ben's Fake Valuations
50	Townhouse	45
20	Red Sox Tickets	25
30	Klee painting	30
100	Total	100

We leave it as an exercise at the end of the section to see that if Annie were honest, Ben and Annie would each get 52 8/11 points, although this would really constitute over 77 points for Ben. But if Annie and Ben both submit these false preferences, the result is not good for either. In the first step of the process, Annie receives the Red Sox tickets, and Ben gets the townhouse and the painting. The ratio for the painting is 1.5 while the ratio for the townhouse is 45/32, strictly less. The following calculation gives the fraction of the townhouse to be given to Annie:

$$48 + 32x = 30 + 45(1 - x) = 75 - 45x$$

$$77x = 27$$

$$x = 27/77$$

So Annie gets just over a third of the townhouse and the Red Sox tickets, while Ben gets just under 2/3 of the townhouse and the painting. Although this appears to be just over 59 points for each with the false point allocations, both Annie and Ben do much worse according to their true preferences. Annie's share gives her $(27/77)(50) + 20$, roughly 37.5 points and Ben's share gives him $(50/77)(30) + 20$, just under 39.5 points. Both Annie and Ben would have fared much better had they been honest!

The adjusted winner procedure then, in addition to guaranteeing an allocation that is envy-free, equitable, and efficient also promotes honesty, at least when knowledge of the other party's preferences is not strictly one-sided.

■ 11.4 FAIR DIVISION PROCEDURES FOR THREE OR MORE PARTIES

In **Section 5.7**, we looked at the adjusted winner procedure, which guarantees an envy-free, equitable, and efficient allocation of goods to both parties. The procedure is for disputes involving two parties only, however, and just as the theory of social choice procedures is much more complicated for three or more alternatives, the theory of fair division procedures is much more complicated for three or more parties.

For our exploration of fair division procedures for three or more parties, we will focus on the case of a single homogeneous item. We continue to use the cake-cutting metaphor as with Austin's procedure in **Section 5.6**. We saw that both the divide-and-choose method and Austin's method provide envy-free solutions for two people, but it is not at all clear if either procedure can be extended to three or more parties. In fact, it wasn't until 1960 that a procedure guaranteeing an envy-free solution for three people was first discovered by John L. Selfridge, and independently by John H. Conway. Since envy-freeness is a stronger property than proportionality for more than two parties, we can also look at allocations that are proportional but not necessarily envy-free. Envy-freeness is of course more desirable, but unfortunately significantly more difficult to satisfy. We'll look at envy-free procedures in **Section 11.4**.

The first procedure we consider was discovered by Hugo Steinhaus in 1948. As Steinhaus discovered, many procedures that work for three people and even procedures that work for four people do not easily extend to the case of five or more. In 1967, however, Harold Kuhn was able to extend Steinhaus's procedure to work for any number of parties.

> **DEFINITION.** Suppose that a cake is to be divided among n parties. We will say that a piece of cake is *acceptable* to a person if it worth at least $1/n$ of the total value of the cake to that person. So an allocation is proportional if and only if each party finds his or her piece of cake acceptable.

The Steinhaus Lone-Divider Procedure

Suppose Annie, Ben, and Chris are dividing a cake amongst themselves.

Step 1: Annie divides the cake into three pieces, each of which she finds acceptable.

Step 2: Ben and Chris each indicate which of the pieces he finds acceptable. Note that each one must find at least one piece acceptable; otherwise, he views each piece as less than one-third of the cake, contradicting the fact that the sum of the three pieces must be 100% of the cake.

Step 3: There are two distinct possibilities: either both Ben and Chris find exactly one of the three pieces acceptable or at least one of the two considers two or more pieces to be acceptable. Note that by the remark above, exactly one of these possibilities must occur.

Step 3A: If both Ben and Chris find only one of the three pieces to be acceptable, then there is at least one of the three pieces that both find unacceptable. Give this piece to Annie. Reassemble the two remaining pieces, and Ben and Chris can split this portion using the two-player divide-and-choose method.

Annie clearly finds her share acceptable since she was the original divider. Ben and Chris agree that the remaining cake is worth at least two-thirds of the total value because they each considered Annie's piece to be strictly less than one-third the cake. The two-player divide-and-choose method then guarantees each a piece worth at least half of the remaining cake, that is, at least half of two-thirds. Thus Ben and Chris each receive an acceptable share of cake.

Step 3B: Suppose at least one of Ben and Chris, say Ben, finds at least two pieces of cake to be acceptable. First, Chris decides which of the three pieces he wants. Next, Ben chooses one of the remaining pieces. Finally, Annie receives the remaining slice.

Since Chris chooses first, he is guaranteed an acceptable piece of cake. Ben finds two pieces acceptable, and at least one of these remains after Chris chooses, so Ben is also guaranteed an acceptable piece of cake. As the original divider, Annie finds all three pieces acceptable, including the one she receives.

The procedure above guarantees each of the three parties a proportional piece of cake, but it is not an envy-free procedure. For example, suppose that Ben and Chris find only one piece acceptable. When the two pieces are reassembled and re-divided according to two-player divide and choose, it might be that Annie considers one to be half the cake and the other one-sixth of the cake. Annie envies the person with half the cake. As another example, suppose that Ben finds two pieces acceptable: he considers one of the two to be one-third of the cake, and one to be one-half the cake. If Chris chooses the piece that Ben thinks is more valuable, then Ben envies Chris.

The Steinhaus method is therefore not an envy-free procedure. We see from the above examples that it is not equitable either. In fact, it is not efficient—it fails all three of our fairness criteria! Suppose the cake has three components: one vanilla, one chocolate, one hazelnut. If Annie likes only vanilla, Ben likes only chocolate, and Chris likes only hazelnut, then the best solution for everyone involved is to give Annie the entire vanilla portion of cake, Ben the chocolate, and Chris the hazelnut. Each receives 100% of the total value of the cake. In the procedure above though, Annie must divide the vanilla portion equally among three pieces of cake in order to guarantee herself an acceptable piece; the resulting allocation will therefore not be as good as the one above. For the same reason, we see that envy-freeness and efficiency are generally difficult to satisfy simultaneously. Any procedure that requires the individual players to cut the cake themselves forces the players to "play it safe" in order to guarantee themselves a proportional share; in the example above, Annie would never put the entire vanilla portion of cake into one share if she is unaware of Ben and Chris's preferences since she risks receiving a share with no vanilla at all. Playing it safe though, as in the example, often prevents the parties from achieving the best possible outcome.

■ 11.5 ENVY-FREE PROCEDURES

It wasn't until 1960 that a procedure guaranteeing an envy-free division of a single heterogeneous item was found by John L. Selfridge and John H. Conway. Surprisingly it was known in the 1940s that if everyone's preferences are "countably additive," then an envy-free allocation definitely exists for any number of people. The proof uses some advanced mathematical analysis, and only proves the existence of such an allocation; it does not provide a method for achieving such an allocation, and so is of little practical value.

The Selfridge-Conway Procedure

Annie, Ben, and Chris are again dividing a cake amongst themselves.

> Step 1: Annie cuts the cake into three pieces that she considers to
> be equally valuable.

Step 2: Ben trims the piece he considers the largest of the three in order to create a tie between the two largest pieces in his opinion. The trimmings are temporarily set aside.

Step 3: Chris chooses the piece he prefers most. Ben next chooses one of the two remaining pieces, with the restriction that he must choose the piece that he trimmed if it remains. Annie receives the last piece.

Before we divide the trimmings, we will show that at this point in the procedure, the allocation is envy-free. Since Chris is the first person to choose a piece, he envies no one. When it is Ben's turn to choose, at least one of the two pieces he considered tied for largest remains, so Ben too envies no one. Now Annie cut the cake into three equal pieces originally, and the trimmed piece was taken by either Chris or Ben. So Annie is guaranteed to receive one of those original pieces which is equal to or greater in size than both Ben's and Chris's, so Annie also envies no one.

One possibility for dividing the trimmings would be to repeat the procedure. Eventually the trimmings would be negligible (quite possibly after only a few iterations in practice). We can achieve an exact envy-free division of the remaining cake, however, with only two additional cuts. Suppose that Chris chose the piece of cake that had been trimmed.

Step 4: Ben cuts the trimmings into three pieces that he considers to be equal in size. If Ben had received the trimmed piece of cake, then Chris cuts the trimmings into three pieces.

Step 5: Chris chooses the piece he prefers most. Annie chooses next, followed by Ben.

Notice that Chris does not envy anyone else since he chose first. Annie does not envy Ben since she chooses before him and therefore prefers her piece to his. To see that Annie does not envy Chris, recall that Chris's first piece was the one that was trimmed. Annie originally divided the cake into what she considered to be three equal pieces. So even if Chris received all of the trimmings, that would give him a piece of cake equal in value to Annie's first piece, excluding her share of the trimmings. Thus Annie does not envy Chris either. Finally, Ben envies no one because he cut the trimmings into three equal pieces, one of which he receives.

The Selfridge-Conway method therefore guarantees each of three parties an envy-free division, although it does require more cuts than the Steinhaus procedure. While this may not be very important for cutting cake, it could raise issues if say land were being divided, since someone might receive two pieces of non-contiguous property. As with the Steinhaus procedure, the Selfridge-Conway method also fails to be efficient and equitable (see exercises).

We saw earlier that efficiency and envy-freeness are difficult to satisfy simultaneously when the parties themselves are doing the cuts. Perhaps a better solution could be obtained by using an alternative method. Another common category of cake-cutting methods involves "moving-knife" procedures. Austin's moving-knife procedure for two parties, which we saw in Chapter 5, is one such method. We now consider a moving-knife procedure for three people, first discovered by William Webb, that uses Austin's method.

The Webb Moving-Knife Procedure

Once again, Annie, Ben, and Chris are dividing a cake. Assume the cake is rectangular.

Step 1: An unbiased fourth party, who unfortunately receives no dessert, slowly moves a knife across the cake until someone yells "Cut!" to indicate that he or she values the piece to be cut off at one-third of the cake. Suppose that Annie is the one who yells "cut," and let P_1, represent the piece of cake that is cut off.

Step 2: Annie and Ben now use Austin's procedure to divide the remaining cake into two pieces that they both consider equally valuable. Let P_2 and P_3 denote these two pieces.

Step 3: Chris chooses first from the three pieces P_1, P_2, and P_3. Ben chooses next, and Annie chooses last.

Chris envies no one since he gets to choose first. Since Annie yelled "cut" the first time, she believes that P_1 is exactly one-third of the cake. She thinks that P_2 and P_3 are equally valuable and together are worth two-thirds of the cake, so she thinks P_2 and P_3 are each exactly one-third of the cake as well. Since she considers each of the three pieces to be equally valuable, she envies no one. Finally, Ben considers P_1 to be less than one-third the cake since he was not the one to yell "cut." So he thinks P_2 and P_3 together make up more than two-thirds of the cake. So Ben values P_2 and P_3 equally, and strictly more than P_1. Since Ben chooses second, at least one of P_2 and P_3 will be available, so he envies no one.

As usual, though, we fail to achieve efficiency, and the Webb procedure is not equitable either. We leave the proof as an exercise.

■ 11.6 ENVY-FREE PROCEDURES FOR FOUR OR MORE PARTIES

In trying to find a method of guaranteeing an envy-free allocation of cake for four parties, it is not unreasonable to attempt to generalize our previous methods for three parties. Suppose Annie, Ben, Chris, and David are the four participants. We could try to generalize the Selfridge-Conway procedure as follows by first having Annie cut the cake into four pieces she values equally. Then Ben and Chris could trim some pieces (we would need to figure out exactly how many each would trim), creating ties for the largest pieces in their estimations. David would choose first, followed by the remaining players in reverse order: Chris, Ben, Annie. Unfortunately, this method is not envy-free, as Annie might envy one of the other three players. We leave the details to the exercises.

It is possible to find an envy-free solution for four parties though. In fact, one can find an envy-free solution for any number of people, although the procedure becomes significantly more complicated. We will content ourselves here with a four-person solution.

Suppose Annie, Ben, Chris, and David are dividing a cake.

Step 1: Annie and Ben use Austin's procedure to divide the cake into two pieces they both consider to be exactly half the cake.

Step 2: Annie and Ben repeat Austin's procedure on each half of the cake. The result is four pieces of cake that Annie and Ben agree are equally valuable.

Step 3: Chris trims one piece (if necessary) to create a tie for the two largest pieces in his opinion. The trimmings are temporarily set aside.

Step 4: David chooses first. Chris chooses next, under the condition that he must choose the trimmed piece if it remains. Annie and Ben choose next, in either order.

At this stage, we have an envy-free allocation of the cake minus the trimmings. David chooses first, so envies no one. Chris is guaranteed to receive one of the two pieces he considers tied for largest so he envies no one. Annie and Ben will each receive a non-trimmed piece of cake, which they considered to be equal in value to the other non-trimmed pieces and strictly better than the trimmed piece; consequently they envy no one either.

Next we distribute the trimmings. Notice that the trimmed piece of cake was chosen by either David or Chris. Call the person with the trimmed piece the "noncutter" and the other the "cutter."

Step 5: Annie and the cutter (hence the name cutter) use Austin's procedure (three times as in Steps 1 and 2) to divide the trimmings into four pieces that each considers equal.

Step 6: From these four pieces, they choose in the following order: noncutter, Ben, Annie, cutter.

The entire cake has now been allocated, and the result is envy-free. The noncutter had first choice of the trimmed pieces, so he envies no one's share of the trimmings. Since he did not envy anyone's original piece either, he does not envy anyone's share of the entire cake. The same reasoning shows that Ben does not envy either Annie or the cutter since he chooses before them. Ben does not envy the noncutter either; of the four original pieces of cake, the noncutter received the trimmed piece. Since Ben received an untrimmed piece, he would consider his piece as valuable as the noncutter's share even if the noncutter received all of the trimmings. Since the noncutter does not receive all the trimmings, Ben strictly prefers his share of cake. Finally, Annie and the cutter consider each of the four shares of trimmings to be equally valuable, so they envy no one's share of trimmings. They did not envy anyone's original piece of cake either, so the entire allocation of cake is envy-free.

Unfortunately, the procedure above does not generalize to more than four parties. It is possible to achieve an envy-free division of cake for any number n of parties, but the procedures are considerably more complicated. Interestingly, the primary difference for the procedures for five or more parties is to cut the cake into more pieces, in the first stage, than there are players. For more on this see Brams and Taylor (1996).

■ 11.7 ANOTHER IMPOSSIBILITY RESULT

We have seen procedures for three or more parties that guarantee envy-freeness, but none were efficient or equitable. In fact, it is impossible to guarantee all three fairness criteria for three or more parties. Again, just as with Arrow's theorem, we emphasize that we are not merely claiming that no procedure guaranteeing all three criteria has yet been discovered; we are claiming that no such procedure will ever be discovered.

We first demonstrate this fact with an example of indivisible goods with three parties and three items. The entry in the first column and i-th row gives the number of points, out of a possible 100, that Annie gives to item i; the second column gives Ben's point valuations, and the third column gives Chris's point valuations. Note that each column adds to 100.

	Annie	Ben	Chris
Item 1	64	46	43
Item 2	30	31	20
Item 3	6	23	37

In this example, it is impossible to distribute the three items in a way that is efficient, equitable, and envy-free. The situation is actually much worse. Notice that each person values item 1 the most, so the two people that do not get item 1 will envy the one who does. Thus no envy-free allocation exists for this example, even if we are willing to sacrifice equitability and efficiency. Since all nine entries in the chart are distinct, there is also no way to achieve an equitable allocation, even if we are willing to sacrifice envy-freeness and efficiency. There are efficient allocations, for example, the one in which Annie gets item 1, Ben gets item 2, and Chris gets item 3. The only way to improve Ben or Chris's outcome would be to give one of them item 1, but this would hurt Annie.

Now with adjusted winner, it was often necessary to divide one item in order to achieve all three fairness criteria. With three or more parties though, even allowing one item to be divided is not sufficient to guarantee that an envy-free, efficient, and equitable allocation exists. In fact, one can show that even allowing *all* items to be divided is not sufficient in the example above.

In the case of indivisible goods which adjusted winner deals with, we can use mixed integer programming to find an envy-free, equitable, and efficient allocation if it exists. If such a solution does not exist, however, a solution satisfying any two of the three can be found. Of course, deciding which of the three criteria to sacrifice is not easy, and may depend on the particular situation. In the example above, we can obtain an efficient and equitable allocation by giving item 2 to Ben, item 3 to Chris, and splitting item 1 approximately in the proportions 0.653, 0.235, 0.112 to Annie, Ben, and Chris, respectively.

■ 11.8 CONCLUSIONS

We began this chapter by proving that the adjusted winner procedure is efficient. Thus, for two parties, it is always possible to achieve an envy-free, equitable, and efficient allocation of goods. Moreover, we saw that unless one party has strictly one-sided information about the preferences of the other party, it pays to be honest.

We next considered cake-cutting methods for three or more parties. As with social choice procedures, with three or more parties, it is significantly more challenging to find an allocation that satisfies all of the desired criteria. We saw one method which guarantees each of three parties a proportional, but not envy-free, piece of cake, and we saw two methods which guarantee each of three or four parties an envy-free piece of cake, though these methods do tend to get quite involved as the number of parties involved grows. Unfortunately, not one of the procedures we considered is efficient or equitable.

We ended the chapter with one last look at the case of indivisible goods. One might hope that the adjusted winner procedure which was so successful for two parties might generalize to three or more parties. This turns out not to be the case, however, and we saw an example with three items and three people where it is impossible to achieve all three fairness criteria.

EXERCISES

1. Prove that after Stage 1 of the adjusted winner procedure (that is, before the equitability adjustment), the division of goods is efficient.

2. Show that, as mentioned in **Section 11.3**, if Annie submits honest point allocations and Ben submits false point allocations as follows, then each person appears to get 52 8/11 points, although Ben truly values his share at over 77 points.

Annie's Sincere Valuations	Item	Ben's Fake Valuations
50	Townhouse	45
20	Red Sox Tickets	25
30	Klee painting	30
100	Total	100

3. Give an example to illustrate that the Selfridge-Conway method for three parties fails to be
   (a) efficient
   (b) equitable

4. Describe a moving-knife procedure that is equivalent to divide-and-choose for two parties.

5. Given the choice between an allocation of cake that is equitable, envy-free, or efficient, which would you choose? Explain thoroughly.

6. Give an example to demonstrate that the Webb moving knife procedure for three parties is not

   (a) efficient
   (b) equitable

7. Suppose that the following represent the sincere point allocations of Annie and Ben.

Annie's Points	Item	Ben's Points
40	House	20
5	Pool table	15
25	Artwork	10
20	Rare book collection	15
10	Boat	40
100	Total	100

   Show that it is possible for either to do better by submitting insincere point allocations if the other party submits sincere point allocations.

8. Annie, Ben, Chris, and David divide a cake amongst themselves according to the following procedure.

   Step 1: Annie cuts the cake into five pieces that she considers equally valuable.

   Step 2: Ben trims up to two of the five pieces to create a three-way tie between the largest three pieces. The trimmings are temporarily set aside.

   Step 3: Chris next trims one of the five pieces if necessary (perhaps further trimming one of the pieces already trimmed by Ben) to create a tie for the two largest pieces. The trimmings are temporarily set aside.

   Step 4: David chooses his favorite of the five pieces. Next Chris chooses one of the largest pieces under the condition that if the piece he trimmed in Step 3 remains, he must choose that piece. Ben chooses one of the largest pieces next under the condition that if a piece that he trimmed in Step 2 remains, he must choose one of those. Annie then chooses one of the untrimmed pieces, at least one of which must remain.

(a) Prove that after Step 4, that is if we disregard the trimmings and the remaining fifth piece left unchosen, no one envies anyone else.

9. Generalize the Selfridge-Conway procedure to four people: Annie, Ben, Chris, and David. Annie cuts the cake into four pieces she values equally. Ben and Chris then trim some pieces (how many?) to create ties for largest. David chooses first, followed by Chris, Ben, and Annie. Fill in the details, and show that Annie may envy one of the other players with this procedure.

10. Consider the following fair division procedure for 3-people: Amy, Beth, and Colin.

- Amy divides the cake into two pieces of equal value in her opinion.
- Beth takes the larger (in her opinion) of the two pieces, and gives the remaining piece to Amy.
- Amy and Beth each divide their piece of cake into three pieces of what they consider to be equal value. There are now six pieces of cake.
- Colin chooses one piece of cake from Amy's three pieces, and one piece of cake from Beth's three pieces. Amy keeps her remaining two pieces and Beth keeps her remaining two pieces.

(a) Is this procedure proportional? Why or why not?

(b) Is this procedure envy-free? Why or why not?

11. We saw that for three or more people, it is not always possible to find an allocation of divisible goods that is equitable, envy-free, and efficient. Which of the three fairness criteria would you most be wiling to sacrifice if you were guaranteed the other two? Explain thoroughly.

12. We saw that for three or more people and indivisible goods, it is not always possible to find an allocation that is efficient, equitable, and envy-free. Suppose that three heirs, Alex, Bella, and Cate, are

dividing their parents' estate. There are 6 items to be assigned: the thoroughbred horse *Old Ironside*, the rare book collection, the original Picasso, the 1957 Chevy Convertible, the house on the Cape, and the Victorian house in Providence. Each of the three heirs allocates a total of 100 points each to the six items. Let $A_1, A_2, \ldots, A_6$ be the points assigned by Alex, $B_1, B_2, \ldots, B_6$ the points assigned by Bella, and $C_1, C_2, \ldots, C_6$ the points assigned by Cate. Let $x_1, x_2, \ldots, x_6$ be the fraction of each item assigned to Alex, $y_1, y_2, \ldots, y_6$, the fraction of each item assigned to Bella, and $z_1, z_2, \ldots, z_6$, the fraction of each item assigned to Cate.

(a)   What is the value of $A_1 + A_2 + \ldots + A_6$? The same value should hold as well for $B_1 + B_2 + \ldots + B_6$ and $C_1 + C_2 + \ldots + C_6$.

(b)   What is the value of $x_1 + y_1 + z_1$? The same value should hold for $x_i + y_i + z_i$ for $i = 2, 3, 4, 5, 6$.

(c)   Write down a set of 6 inequalities that hold if and only if the allocation is envy-free.

(d)   Write down a set of equalities that hold if and only if the allocation is equitable.

13.   Eighteen cookies are to be divided between three good friends (Michael, Mike, and Peter) after a hard night's work in Athens, Georgia. There are six chocolate chip cookies, 6 peanut butter cookies, and 6 sugar cookies with rainbow sprinkles. Michael is thinking of going vegan (he's already a vegetarian), so the chocolate chip cookies are worthless to him (fortunately, the peanut butter and sugar cookies were made without eggs, butter, or milk). He likes the peanut butter and sugar cookies equally.

Mike is allergic to peanuts, so he cannot eat the peanut butter cookies. He likes the chocolate chip and sugar cookies equally. Peter likes the chocolate chip and peanut butter cookies equally but does not like the sugar cookies at all—the sprinkles fall into his mandolin. Give examples of allocations of cookies (all 18 must be accounted for) that are

(a)   envy-free but not equitable

(b)   equitable but not envy-free

14.   Suppose that Annie, Ben, Chris, and David are dividing a collection of cupcakes among them. There are four varieties: lemon meringue cupcakes, peanut butter cupcakes with chocolate frosting, carrot cake cupcakes, and chocolate espresso cupcakes. Annie detests

carrot cake and considers those cupcakes worthless. She likes the lemon meringue and peanut butter cupcakes equally, but thinks the chocolate espresso are twice as good. Ben is allergic to peanuts, so the peanut butter cupcakes are worthless to him, but he likes the other flavors equally. Chris tries to avoid caffeine, so while he likes the chocolate espresso cupcakes, he thinks the lemon meringue are twice as good. He thinks the lemon meringue and peanut butter are equal in value, and three times as good as the espresso. Finally, David likes all cupcakes equally, and just wants as many as possible.

Show that the envy-free procedure described in **Section 11.6** can be used in this setting to find an envy-free allocation of cupcakes.

15. The following example is due to J.H. Reijnierse and J.A.M. Potters.

	Annie	Ben	Chris
Item 1	40	30	30
Item 2	50	40	30
Item 3	10	30	40

Prove that if Annie, Ben, and Chris distribute their points as shown, then it is impossible to achieve all three fairness criteria without dividing any of the items.

CHAPTER
12

More Escalation

▮ 12.1 INTRODUCTION

This chapter contains a proof of a very pretty theorem due to Barry O'Neill that prescribes optimal play (for rational bidders) in the dollar auction from Chapter 6. We prove this theorem in **Sections 12.2** and **12.3**. In **Section 12.4**, we explain the sense in which a Vickrey auction is a "generalized Prisoner's Dilemma."

▮ 12.2 STATEMENT OF THE STRONG VERSION OF O'NEILL'S THEOREM

To facilitate the discussion in this section and the next, we shall work with the special case of the dollar auction in which $s = 20$ and $b = 100$. If the units are nickels, this corresponds to the stakes being one dollar and the bankroll five dollars. We also assume throughout this section and the next that we are working with the setup as described in Chapter 6: the second high bidder pays whatever he or she bid and receives nothing, both bidders use the conservative convention, etc.

Recall that O'Neill's theorem tells us to calculate the optimal opening bid (for bankroll b and states s) by subtracting $s - 1$ from b repeatedly until we reach the point where one more subtraction would make the result zero or less than zero. This point reached is the optimal opening bid. If $b = 100$ and $s = 20$, this means subtracting 19 repeatedly from 100 to obtain the sequence

$$100, 81, 62, 43, 24, 5$$

at which point we stop since one more subtraction would yield -14, which is less than zero. Hence, according to O'Neill's theorem, the optimal opening bid in this case is 5 units, which, if the units are nickels, is twenty-five cents.

The statement of what we shall call the "strong version of O'Neill's theorem" requires a bit of terminology. Let's agree to call the bids 5, 24, 43, 62, 81, and 100 *special numbers*. Thus, the rational opening bid, according to O'Neill's theorem, is the smallest special number. The sense in which the rest of these numbers are "special" will be made clear in a moment.

Let's also say that a bid is *rationally unavailable* to a bidder if it exceeds his or her previous bid (or zero, if no one has bid yet) by 20 or more. The choice of terminology is explained by the following.

LEMMA A. *In the dollar auction with the conservative convention and $s = 20$, it is never rational to choose a bid that exceeds your previous bid (or zero, if no one has bid yet) by 20 or more.*

PROOF. See Exercise 1. (This exercise also occurred at the end of Chapter 6.)

If a bid is not rationally unavailable, we will say that it is *rationally available*. Notice, however, that a rationally available bid is not necessarily a rational choice for a bid; it is simply not irrational for the particular reason we are discussing (i.e., exceeding your previous bid by 20 or more).

We shall need one more lemma in the proof of O'Neill's theorem. We record it here.

LEMMA B. *In the dollar auction with the conservative convention, suppose that there is a legal bid that*

1. *is rationally available to you (that is, it exceeds your last bid by at most 19), and*

2. *will result in a pass by your opponent.*

Then making this bid yields a strictly better outcome for you than does passing or bidding higher than this. (Thus, while we cannot say that making this bid is definitely the rational thing to do, we can say that neither passing nor bidding higher than this is rational.)

PROOF. See Exercise 2.

The proof of O'Neill's theorem involves doing something that should strike the reader as slightly strange: verifying the truth of an assertion that is strictly stronger than the theorem we are trying to prove. Moreover, this is not something we are doing because we think the stronger assertion is worth any extra effort it might entail (although we would argue that it does). The point is that the proof technique we shall be using (known as mathematical induction) often requires proving a stronger statement than the assertion in which one is really interested.

The stronger version of O'Neill's theorem that we consider provides an answer to the following question: Suppose two irrational people (Bob and Carol) have been engaged in the dollar auction and suppose Bob made the last bid, which we shall denote by the letter x. Suppose now that Bob wants you to take over for him (even though x may well exceed the stakes at this point), but he and Carol at least provide you with the guarantee that Carol will be rational from now on (and that she knows you are rational, etc.). Should you pass or bid? If you choose to bid, how much?

In terms of the game-tree analyses we did in Chapter 6, the question in the previous paragraph is equivalent to asking the following: Given a node n on the tree, suppose we erase everything except that part of the tree below the node n and connected to it by a branch. What is left is again a tree, and it can be "pruned" as before. Which of the nodes immediately below and connected to the node n survives the pruning process? The node n (which becomes the top of the new tree) corresponds to the "state of the auction" when Bob turns it over to you; the surviving node tells you what (if anything) to bid. (Exercises 3 and 4 provide specific examples of this kind of game-tree analysis.)

The answers to the questions in the two previous paragraphs are provided by the following:

THEOREM (*Strong Version of O'Neill's Theorem*). *Suppose you are engaged in the dollar auction with the stakes $s = 20$, the bankroll $b = 100$, and the conservative convention. Assume that your opponent's last bid was x (even though bidding x may have been an irrational thing for him or her to have done). Assume that from this point on, your opponent will definitely be rational, knows you will be rational, knows that you know that he or she will be rational, etc. Then the rational course of action is for you to*

1. *bid the smallest special number that is greater than x if one exists and is rationally available to you, and*

2. *pass otherwise.*

COROLLARY (*O'Neill's Theorem*). *The optimal opening bid in the dollar auction with the conservative convention is the smallest special number.*

Our decision to work with the special case $s = 20$ and $b = 100$ was based on a desire to simplify the proof. The statement of the general version is an easy variant of the above (see Exercise 5).

Notice that the conclusion of the strong version of O'Neill's theorem is really a sequence of 100 statements (that we choose to number from 0 to 99 instead of from 1 to 100):

Statement 0: If $x = 100$, then you should pass.
Statement 1: If $x = 99$, then you should bid 100 if it is rationally available, and pass otherwise.
$\vdots$

Statement 37: if $x = 63$, then you should bid 81 if it is rationally available, and pass otherwise.

$\vdots$

Statement 99: If $x = 1$, then you should bid 5 if it is rationally available, and pass otherwise.

It might seem more natural to have reversed the ordering of the statements so that Statement 1 (or 0) would correspond to $x = 0$; the reason we chose not to do so will be clear in the next section.

The proof technique used in establishing the strong version of O'Neill's theorem is called *mathematical induction*. Our choice to work with the special case $b = 100$ and $s = 20$ in the theorem was made so that we could focus on a concrete illustration of the "ladder climbing" idea (illustrated in what follows) that underlies this technique of proof.

To prove the strong version of O'Neill's theorem, we must verify that every one of the 100 statements in its conclusion is true. Notice that Statement 0 (corresponding to $x = 100$) is trivial to verify: The theorem says you should pass in this case, and you have no choice but to do so since your opponent just bid the whole bankroll. At the other extreme, if we had a Statement 100 (corresponding to $x = 0$) it would literally be O'Neill's theorem as stated in Chapter 1, and this is, after all, what we are primarily interested in. In fact, our choice to have O'Neill's theorem stated as a corollary instead of built into the theorem as Statement 100 was based entirely on a desire to avoid having x play two different roles in the proof depending upon whether we were talking about the first bid or not. Exercise 6 asks the reader to recast the theorem and proof to avoid needing the corollary.

Thus, in some sense, the statements become harder to verify as x moves down from 100 to 99 to 98 and on down to 1. So, if we start by verifying Statement 0 ($x = 100$, which we have already done), and then move to Statement 1 ($x = 99$), and then to Statement 2 ($x = 98$), and so on, what is it that will allow us to continue to succeed in proving the statements are true as they become harder and harder to verify?

The answer is that when we try to verify the statement corresponding to, say, $x = 97$, we will have something more to work with than if we tried to prove it right now. The "something more" we will have is the knowledge that the statements corresponding to $x = 100$, $x = 99$, and $x = 98$ are definitely true and thus can be used in our verification of the statement corresponding to $x = 97$. This is the key idea behind mathematical induction.

To see how this goes, we will start with $x = 100$ and verify the first few assertions. (If we were to provide a separate argument for each of the 100 statements, this would be a very long proof. Fear not.)

Statement 0 (x = 100)

We have already verified this one, and it turned out to be completely trivial: Given the fact that our opponent has just bid the whole bankroll, we have no choice but to pass, and this is what the theorem prescribes since no special number is available to us as a bid under the rules of the auction.

Statement 1 (x = 99)

There are only two available options—pass or bid 100. Notice also that 100 is a special number, and so the theorem says we should bid it if it is rationally available to us, and pass otherwise. But if 100 is rationally available to us, then Lemma B says we should not pass (since we know a bid of 100 will result in pass by our opponent). Thus, we should bid 100 in this case as the theorem says. On the other hand, if 100 is rationally unavailable to us, then we should not bid it (by Lemma A), and so we should pass in this case, which is once again what the theorem says.

Statement 2 (x = 98)

CLAIM. Passing is better than bidding 99.

> **PROOF.** Notice that 100 is certainly rationally available to our opponent since it exceeds his or her previous bid by only 2. Thus, if we bid 99, then we know, because of the truth of Statement 1 (which is the $x = 99$ case), that our opponent will bid 100. Thus, bidding 99 will result in our losing 99, and this is strictly worse for us than passing.

Hence, we should either pass or bid 100. Lemma B now says that we should bid 100 unless it is rationally unavailable to us, in which case we should pass, by Lemma A. Thus, we have verified the theorem in this case.

Statement 3 ($x = 97$)

CLAIM. Passing is better than bidding 99 or 98.

PROOF. Notice that 100 is certainly rationally available to our opponent since it exceeds his or her previous bid by only 3. Thus, if we bid 99 or 98, then we know, because of the truth of Statements 1 and 2 (which are the $x = 99$ and $x = 98$ case), that our opponent will bid 100. Thus, bidding 99 or 98 will result in our losing 99 or 98, and this is strictly worse for us than passing.

Hence, we should either pass or bid 100. Lemma B now says that we should bid 100 unless it is rationally unavailable to us, in which case we should pass by Lemma A. Thus, we have verified the theorem in this case.

Statements 4 ($x = 96$) Through 19 ($x = 81$)

The arguments for $x = 96, \ldots, 81$ are analogous to what we just did because, in all these cases, 100 is rationally available to our opponent. (Exercise 7 asks for the case $x = 96$.) Thus, we can assume we have verified the statement for $x = 81$ through $x = 100$. Notice also that if $x = 81$, then 100 is rationally unavailable to us, and so we should pass. Let's do one more.

Statement 20 ($x = 80$)

If 81 (which is the smallest rational number greater than x) is rationally unavailable to us, then so is every bid greater than 81 and so we should pass, by Lemma A. If 81 is rationally available to us, then a bid of it will result in a pass by our opponent, since 100 (the next largest special number) is rationally unavailable to him since it exceeds his last bid of 80 by 20, and we know that Statement 19 ($x = 81$) is true. Lemma B now guarantees that 81 is the rational bid for us to make in this case.

At this point, the reader is probably convinced that we could present a sequence of 100 proofs—each one built upon the facts established by the preceding ones—verifying all 100 statements corresponding to the 100 possible previous bids by our opponent. What we need, of course, is an argument that convinces us of the existence of the sequence of 100 proofs without actually producing each one of them. This is precisely what a general proof by mathematical induction does. We illustrate this for the theorem at hand in the following section.

■ 12.3 PROOF (BY MATHEMATICAL INDUCTION) OF THE STRONG VERSION OF O'NEILL'S THEOREM

For $n = 0$ to $n = 99$, let "Statement n" be the following assertion:

> Statement n: If your opponent's last bid was $x = 100 - n$ and it is now your bid, then the rational course of action is for you to bid the smallest special number that is greater than x if this bid is rationally available to you, and to pass otherwise.

Notice first that Statement 0 (corresponding to the state of the auction being $x = 100$) is certainly true, since the rules of the auction force you to pass at this point and passing is what the statement prescribes in this case. (Recall also that we are assuming that everyone is rational from this point in the auction on.)

For the so-called inductive step, assume that n is an arbitrary number between 0 and 98 and that we have verified the truth of Statement 0 through Statement n. This assumption of the truth of Statements 0 through n is called the *inductive hypothesis*. We want to show that Statement $n + 1$ is also true. Thus, we are assuming that

1. your opponent's last bid was $x = 100 - n$,

2. it is now your bid, and

3. we know what rational play will yield when *anyone's* last bid is between x and 100.

Notice for 3 that, although we are phrasing our arguments as if they are applying to the reader, the conclusions we reach surely apply equally well to the reader's opponent.

Let La denote our last bid and let Sp denote the smallest special number greater than x. Thus:

$$La < x < Sp.$$

Notice that Sp is rationally available to our opponent since it is the smallest special number greater than x and so exceeds x by at most 19. Thus, if we bid anything from $x + 1$ to $Sp - 1$, we know from the inductive hypothesis that our opponent will bid Sp and then we will pass. Thus, bidding anything from $x + 1$ to $Sp - 1$ is worse for us than passing and so is not rational.

On the other hand, bidding Sp will—because of the inductive hypothesis again—definitely result in a pass by our opponent, since the next special number is rationally unavailable to him or her. This shows that bidding anything greater than Sp is not rational (since it is worse for us than a bid of Sp).

Thus, the only candidates for a rational bid left for us are passing and bidding Sp. Lemmas A and B now yield the desired result: If Sp is rationally unavailable to us, we should pass by Lemma A. If Sp is rationally available to us, then (since we know it will force a pass by our opponent) we should not pass, by Lemma B. This completes the proof.

Proof of the Corollary (O'Neill's Theorem).

The theorem guarantees us that if we make an opening bid less than the smallest special number Sp, our opponent will respond with Sp as a bid and we will pass, thus losing money. On the other hand, if we open with Sp as a bid, then our opponent will pass (since the next larger special number will be rationally unavailable to him or her). Winning with a bid of Sp is better for us than winning with a larger opening bid (or losing). Hence, we have shown that opening with a bid of Sp is the rational course of action, as desired.

■ 12.4 VICKREY AUCTIONS AS A GENERALIZED PRISONER'S DILEMMA

Rothkopf, Teisberg, and Kahn (1990) discuss several reasons why Vickrey auctions are seldom used in the real world. Five of these reasons are dismissed by the authors and two are supported: (1) concerns over bidding cheating, and (2) bidder reluctance (based on a consideration of future transactions) to use truth-revealing strategies. In this section, we offer a more structural explanation, based on Taylor–Zwicker (1995c), of the scarcity of Vickrey auctions. This explanation is based on the observation that a Vickrey auction is a generalized Prisoner's Dilemma.

Recall from Chapter 4 that Prisoner's Dilemma (PD) is the following 2×2-ordinal game:

		Column	
		C	N
Row	C	(3, 3)	(1, 4)
	N	(4, 1)	(2, 2)

A strategy in this case is a choice of "C" or "N". Strategies are often denoted by the Greek letters σ and τ. Thus, we might say: "Consider the strategy $\sigma = $ C for the Row player." If we were talking about auctions instead of 2×2 ordinal games, a strategy would be a choice of what to bid. Thus, we might say: "Consider the strategy σ that corresponds to bidding one's true valuation of the worth of the object being sold."

The thing that makes PD paradoxical is that each player has a dominant strategy of N, but the resulting (2, 2) outcome is worse for both players than the (3, 3) result arrived at if both were to use the strategy C. Thus, we might say that the strategy sequence $\langle$C, C$\rangle$ "dominates" the strategy sequence $\langle$N, N$\rangle$ in the sense that the use of $\langle$C, C$\rangle$ yields a better outcome for both players than the use of $\langle$N, N$\rangle$. Notice here that we are talking about the use of a *sequence* of strategies, meaning that Player 1 (Row) uses the first strategy in the sequence and Player 2 (Column) use the second strategy in the sequence.

To speak of generalized versions of PD, we need to generalize the notions from the previous paragraph. We know what it means to say that a strategy σ dominates a strategy τ: σ yields a strictly better outcome for the player using it than does τ. For our purposes, though, we are more interested in the notion of a strategy σ weakly dominating a strategy τ, which, we recall, means that σ yields an outcome at least as good for the player using it as does τ in every scenario, and there is at least one scenario in which σ yields a strictly better outcome for that player than does τ.

Turning now to sequences of strategies in an n-person game, let's say that the *sequence* $\langle \tau_1, \ldots, \tau_n \rangle$ of *strategies weakly dominates* the *sequence* $\langle \sigma_1, \ldots, \sigma_n \rangle$ of strategies if and only if:

1. the outcome resulting from the players use of the τ's is at least as good for every player as the outcome resulting from the players use of the σ's, and

2. the outcome resulting from the players use of the τ's is strictly better for at least one player than the outcome resulting from the players use of the σ's.

With this terminology at hand, we can now offer the following:

DEFINITION. An *n*-person game will be called *a generalized Prisoner's Dilemma* if and only if there are two sequences of strategies,

$$\langle \sigma_1, \ldots, \sigma_n \rangle \text{ and } \langle \tau_1, \ldots, \tau_n \rangle$$

such that

1. for every *i*, the strategy σ_i is the unique strategy for Player *i* that weakly dominates every other strategy for Player *i*, and

2. the sequence $\langle \tau_1, \ldots, \tau_n \rangle$ weakly dominates the sequence $\langle \sigma_1, \ldots, \sigma_n \rangle$.

Where this is all heading is given by the following:

PROPOSITION. *A Vickrey auction (in which the players are the n bidders, but not the bid-taker) is a generalized Prisoner's Dilemma.*

PROOF. For each i let σ_i be the truth-revealing strategy for Player i. Then we know from the theorem in the last section that σ_i weakly dominates every other strategy that is available to Player i. However if τ_i is the strategy in which Player i bids (say) exactly half of what he or she thinks the object is really worth, then the sequence $\langle \sigma_1, \ldots, \sigma_n \rangle$ is weakly dominated by the sequence $\langle \tau_1, \ldots, \tau_n \rangle$. That is, for every bidder except the high bidder, the outcome resulting from the players' use of the τ's is the same as the outcome resulting from the players' use of the σ's (i.e., a loss of the auction), while, for the high bidder, the sequence $\langle \tau_1, \ldots, \tau_n \rangle$ yields a strictly better outcome than the sequence $\langle \sigma_1, \ldots, \sigma_n \rangle$ since he or she pays only half as much for the object won.

The above theorem suggests why bidders might be reluctant to bid honestly in a Vickrey auction. For example, suppose we have six contractors bidding for a job. Then, even if all six can be convinced that they can individually do no better than to make honest bids, it is both true and relatively transparent that cooperation benefits all six in the long run. Moreover, the phrase *in the long run* is quite appropriate here, since in many such situations it is the same collection of six contractors who will be engaged in repeated play of this game (and repeated play is well known to produce cooperation in PD). Moreover, in good economic times, bidders such as these six contractors have even more incentive to cooperate, since losing one auction simply means a short delay before it is their "turn" to win.

■ 12.5 CONCLUSIONS

In this chapter we proved Barry O'Neill's theorem giving the optimal opening bid in the dollar auction. We also illustrated the sense in which a Vickrey auction is a generalized prisoner's dilemma.

EXERCISES

1. (a) Show that in the dollar auction, it is never rational to increase your previous bid by more than s units. (Do not use O'Neill's theorem.)

   (b) Show that in the dollar auction with the conservative convention, it is never rational to increase your previous bid by more than $s - 1$ units. (Do not use O'Neill's theorem.)

2. Prove Lemma B from **Section 12.2**.

3. In the dollar auction with $b = 6$ and $s = 3$, assume that you (Player 1) open with a bid of 2 as O'Neill's theorem says, but your opponent irrationally responds with a bid of 3. Assume, however, that from now on your opponent will be rational, follow the conservative convention, etc.

   (a) Draw the part of the game tree for $b = 6$ and $s = 3$ that has a bid of 3 by Player 2 in place of the start node. (The tree will have eight terminal nodes.)

   (b) Do the game tree analysis (backward induction) as in Chapter 6 to show that your optimal response is now to bid 4 as prescribed by the strong version of O'Neill's theorem.

4. Redo Exercise 3 under the assumption that your opponent responded to your opening bid of 2 with a bid of 4.

5. Define "special bid" for the general case where the stakes are s and the bankroll is b in O'Neill's theorem, and then state the corresponding strong version of O'Neill's theorem.

6. Restate the strong version of O'Neill's theorem so that it involves 101 statements instead of 100 and so that Statement 100 is O'Neill's theorem. (This requires slightly changing the role of x.)

7. Write out the verification of the $x = 96$ case in the proof of the strong version of O'Neill's theorem.

Attributions

Chapter 1

The inspiration for much of what is here is Straffin's early work (1977),
although he did not consider independence of irrelevant alternatives
in this context. (We did so to make Arrow's theorem in Chapter 7 a
little easier.) As mentioned in the preface, the student team of Eskin,
Johnson, Powers, and Rinaldi contributed to the exercises. The impossibility theorem, as far as we know, originated in the first edition of
this book.

Chapter 2

Most of the material in Chapter 2 is based on joint work of Alan
Taylor and William Zwicker, much of which is included in Taylor and
Zwicker (1999). The nonweightedness of the U.N. Security Council is
well known. The Canadian System was first analyzed by Marc Kilgour
(1983) and is also treated in Straffin (1993). The characterization

theorem is due to Taylor and Zwicker (1992), although an early version in a different context (and via a very different proof) goes back to Elgot (1960).

■ Chapter 3

The material on the Shapley-Shubik index, the European Economic Community, and the paradox of new members is well known—see Brams (1975), for example. More on the Banzhaf index can be found in Dubey and Shapley (1979).

■ Chapter 4

Dominant strategies and Nash equilibria are standard fare, as is our discussion of Prisoner's Dilemma and its application as a model of the arms race. The game of Chicken is well known, as is its use in modeling the Cuban missile crisis (although the second model given was new to the first edition of this book). The material on the Yom Kippur War and the theory of moves is from Brams (1994), although our rules are slightly different. The OPEC example in the exercises, and the treatment of Newcomb's program there both originated in the first edition of this book, although others, including Steven Brams, have noticed that there are games in which a player has a dominant strategy that should not be used if he or she is moving first. The game-theoretic analysis of *Tosca* goes back to Rapoport (1962) and Hypergame is due to Zwicker (1987).

■ Chapter 5

The material on apportionment is standard fare, while the example in Section 5.4 is a slight modification of a known result. The adjusted winner procedure is due to Steven Brams and Alan Taylor.

■ Chapter 6

The dollar auction is attributed to Martin Shubik (1971). Our description (and the use of the conservative convention) is based on the article by Barry O'Neill (1986). The game-tree analyses and the section on limitations use well-known techniques, but their application in this context may be new. More on the dollar auction can be found in Leininger (1989).

■ Chapter 7

The version of May's theorem given is a slight elaboration of the usual one. For more on this, see Young, Taylor, and Zwicker (1995). Our proofs of Arrow's theorem and Sen's theorem are based on well-known ideas, but differ from the usual in what we think are some significant ways. For more on Arrow's Theorem, see Kelly (1978), and for a deeper treatment of Sen's theorem, see Zwicker (1991). The proof of the Gibbard-Satterthwaite theorem is from Taylor (2002, 2005).

■ Chapter 8

Virtually all of the material in this chapter is due to Taylor and Zwicker and can be found in Taylor and Zwicker (1993, 1999). Results similar to the theorem in Section 8.2, but from quite a different context, were obtained by Gableman (1961) in his Ph.D. thesis.

■ Chapter 9

Most of what is in Chapter 9 is based on well-known material cited there. For more on ordinal notions of power, see Taylor and Zwicker (1999). The theorem on voting blocs is due to Straffin (1977, 1980).

■ Chapter 10

The models of deterrence are somewhat standard fare, as is the material on two-person zero-sum games. The probabilistic model of deterrence is due to Brams (1985a). The idea in exercises 3 and 4 is not ours.

■ Chapter 11

The four-person moving-knife scheme in Section 11.6 is due to Brams, Taylor, and Zwicker. The fact that AW does not extend to three parties is due to J.H. Riejnierse and J.A.M. Potters. For more on what is treated here, see Brams and Taylor (1996).

■ Chapter 12

The proof we give of O'Neill's theorem originated in the first edition, but it uses the same ideas as in O'Neill's original proof (1986). The Vickrey auction material is well known, although viewing it as a generalized Prisoner's Dilemma is from Taylor and Zwicker (1995).

References

Allingham, Michael. (1975). Economic power and values of games. *Z. Nationalökonomie* 35:293–299.

Arrow, Kenneth. (1950). A difficulty in the concept of social welfare. *Journal of Political Economy* 58:328–346.

Austin, A.K. (1982). Sharing a cake, *Mathematical Gazette*. 66:212–15.

Balinski, Michel and H. Peyton Young. (1982). *Fair Representation: Meeting the Ideal of One Man, One Vote*. New Haven, CT: Yale University Press.

Banzhaf, John. (1965). Weighted voting doesn't work: a mathematical analysis. *Rutgers Law Review* 19:317–343.

Barbanel, Julius. (2005). *The Geometry of Efficient Fair Division*. Cambridge, England: Cambridge University Press.

Black, Duncan. (1958). *Theory of Committees and Elections*. Cambridge: Cambridge University Press.

Brams, Steven. (1975). *Game Theory and Politics*. New York: Free Press.

Brams, Steven. (1983). *Superior Beings: If They Exist, How Would We Know? Game-Theoretic Implications of Omniscience, Omnipotence, Immortality, and Incomprehensibility*. New York: Springer-Verlag.

Brams, Steven, and Peter Fishburn. (1983). *Approval Voting*. Cambridge, MA: Birkhäuser Boston.

Brams, Steven. (1985a). *Rational Politics: Decisions, Games, and Strategy*. Washington, DC: CQ Press.

Brams, Steven. (1985b). *Superpower Games: Applying Game Theory to Superpower Conflict*. New Haven, CT: Yale University Press.

Brams, Steven. (1990). *Negotiation Games: Applying Game Theory to Bargaining and Arbitration*. New York: Routledge.

Brams, Steven. (1994). *Theory of Moves*. Cambridge: Cambridge University Press.

Brams, Steven, Paul Affuso, and D. Marc Kilgour. (1989). Presidential power: a game-theoretic analysis, in *The Presidency in American Politics* (P. Bruce, C. Harrington, and G. King, eds.) New York: N.Y.U. Press, 55-74.

Brams, Steven and Donald Wittman. (1981). Nonmyopic equilibria in 2×2 games. *Conflict Management and Peace Science* 6:39–62.

Brams, Steven and Alan Taylor. (1995). An envy-free cake division protocol. *American Mathematical Monthly* 102:9–18.

Brams, Steven and Alan Taylor. (1996). *Fair Division: From Cake-Cutting to Dispute Resolution*. Cambridge, England: Cambridge University Press.

Brams, Steven and Alan Taylor. (1999). *The Win-Win Solution: Guaranteeing Fair Shares to Everybody*. New York: Norton.

Brams, Steven, Alan Taylor, and William Zwicker. (1995). Old and new moving-knife schemes. *Mathematical Intelligencer* 7:30–35.

Brams, Steven, Alan Taylor, and William Zwicker. (1997). A moving-knife solution to the four-person envy-free cake-division problem, *Proceedings of the American Mathematical Society* 125:547–554.

Caroll, Maureen T., Elyn K. Rykken, and Jody M. Sorensen, The Canadians Should Have Won!?, *Math Horizons*, February 2003.

Dahl, Robert. (1957). The concept of power. *Behavioral Sci.* 2:201–215.

Deegan, J. and Edward Packel. (1978). A new index for simple n-person games. *International Journal of Game Theory* 7:113–123.

Dubey, Pradeep and Lloyd Shapley. (1979). Mathematical properties of the Banzhaf power index. *Mathematics of Operations Research* 4:99–131.

Elgot, C.C. (1960). Truth functions realizable by single threshold organs. AIEE Conference Paper 60–1311.

Felsenthal, Dan and Moshé Machover. (1994). Postulates and paradoxes of relative voting power–a critical re-appraisal. *Theory and Decision*, 38:195–229.

Felsenthal, Dan and Moshé Machover. (1998). *The Measurement of Voting Power: Theory and Practice*. Cheltenham, U.K.: Edward Elgar.

Felsenthal, Dan, Moshé Machover, and William Zwicker (1998). The bicameral postulates and indices of a priori voting power. *Theory and Decision* 44:83–116.

Fishburn, Peter. (1973). *The Theory of Social Choice*. Princeton: Princeton University Press.

Gableman, Irving. (1961). The functional behavior of majority (threshold) elements, Ph.D. thesis, Syracuse University.

Gibbard, Allan. (1973). Manipulation of voting schemes: a general result. *Econometrica* 41:587–601.

Johnston, R.J. (1978). On the measurement of power: some reactions to Laver. *Environment and Planning*. 10A:907–914.

Kelly, Jerry. (1978). *Arrow Impossibility Theorems*. New York: Academic Press.

Kilgour, D. Marc. (1983). A formal analysis of the amending formula of Canada's Constitution Act. *Canadian Journal of Political Science* 16:771–777.

Leininger, Wolfgang. (1989). Escalation and cooperation in conflict situations: the dollar auction revisited. *Journal of Conflict Resolution* 33:231–254.

Massoud, T.G. (2000). Fair Division, Adjusted Winner Procedure (AW), and the Israeli-Palestinian Conflict, *Journal of Conflict Resolution* 44:333–358.

May, Kenneth. (1952). A set of independent, necessary and sufficient conditions for simple majority decision. *Econometrica* 20: 680–684.

McAfree, Preston and John McMillian. (1987). Auctions and bidding. *Journal of Economic Literature* 25:699–738.

Milgrom, Paul. (1989). Auctions and bidding: a primer. *Journal of Economic Perspectives* 3:3–22.

Nurmi, Hannu. (1987). *Comparing Voting Systems*. Dordrecht, Holland: D. Reidel Publishing Company.

O'Neill Barry. (1986). International escalation and the dollar auction. *Journal of Conflict Resolution*. 30:33–50.

Packel, Edward. (1981). *The Mathematics of Gambling and Gaming*. New Mathematical Library #28, The Mathematical Association of America.

Rapoport, Anatol. (1962). The use and misuse of game theory. *Scientific American* 207:108–118.

Robinson, Julia. (1951). An iterative method of solving a game. *Annals of Mathematics* 54:296–301.

Rothkopf, M., T. Teisberg, and E. Kahn. (1990). Why are Vickrey auctions rare? *Journal of Political Economy* 98:94–109.

Saari, Donald. (2001). *Chaotic Elections: A Mathematician Looks at Voting*. The American Mathematical Society.

Saari, Donald. (2001). *Decisions and Elections: Explaining the Unexpected*. Cambridge University Press.

Satterthwaite, Mark. (1975). Strategy-proofness and Arrow's conditions: existence and correspondence theorems for voting procedures and social welfare functions. *Journal of Economic Theory* 10:187–217.

Sen, Amartya. (1966). A possibility theorem on majority decision. *Econometrica* 34:491–496.

Shapley, Lloyd and Martin Shubik. (1954). A method for evaluating the distribution of power in a committee system. *American Political Science Review* 48:787–792.

Shubik, Martin. (1971). The dollar auction game: a paradox in nonco-operative behavior and escalation. *Journal of Conflict Resolution* 15:545–547.

Straffin, Philip. (1977). The power of voting blocs: an example. *Mathematics Magazine* 50:22–24.

Straffin, Philip. (1980). *Topics in the Theory of Voting*. Boston: Birkhauser.

Straffin, Philip. (1993). *Game Theory and Strategy*. The Mathematical Association of America.

Taylor, Alan. (1997). A glimpse of impossibility, *Perspectives on Political Science* 26:23–26.

Taylor, Alan. (2002). The manipulability of voting systems, *American Mathematical Monthly* 109:321–337.

Taylor, Alan. (2005). A paradoxical Pareto frontier in the cake-cutting context, *Mathematical Social Sciences* 50:227–233.

Taylor, Alan. (2005). *Social Choice and the Mathematics of Manipulation*, Cambridge, England: Cambridge University Press.

Taylor, Alan and William Zwicker. (1992). A characterization of weighted voting. *Proceedings of the American Mathematical Society* 115:1089–1094.

Taylor, Alan and William Zwicker. (1993). Weighted voting, multicameral representation, and power. *Games and Economic Behavior* 5:170–181.

Taylor, Alan and William Zwicker. (1995). Vickrey auctions: a generalized Prisoner's Dilemma, preprint.

Taylor, Alan and William Zwicker. (1995b). Simple games and magic squares. *Journal of Combinatorial Theory* (A) 71:67–88.

Taylor, Alan and William Zwicker. (1996). Quasi-weightings, trading, and desirability relations. *Games and Economic Behavior* 16: 331–346.

Taylor, Alan and William Zwicker. (1997). Interval measures of power. *Mathematical Social Sciences* 33:23–74.

Taylor, Alan and William Zwicker. (1999). *Simple Games: Desirability Relations, Trading, and Pseudoweightings*. Princeton: Princeton University Press.

Vickrey, William. (1961). Counterspeculation, auctions, and competitive sealed tenders. *Journal of Finance* 16:8–37.

von Neumann, John. (1928). Zur Theorie der Gesellschaftsspiele. *Mathematische Annalen* 100:295–320. English translation in R.D. Luce and A.W. Tucker, eds. *Contributions to the Theory of Games IV* (1959), pp. 13–42. Princeton: Princeton University Press.

Young, Steven, Alan Taylor, and William Zwicker. (1995). Social choice and the Catalan numbers, *Mathematics Magazine* 68:331–342.

Zagare, Frank. (1987). *The Dynamics of Deterrence*. Chicago: University of Chicago Press.

Zwicker, William. (1987). Playing games with games: the hypergame paradox. *American Mathematical Monthly* 94:507–514.

Zwicker, William. (1991). The voters' paradox, spin, and the Borda count. *Mathematical Social Sciences* 22:187–227.

Index